T0241883

Lecture Notes in Geosystems Mathematics and Computing

Series Editors

Willi Freeden
M. Zuhair Nashed
O. Scherzer

More information about this series at http://birkhauser-science.com/series/13390

Murat Uzunca

Adaptive Discontinuous Galerkin Methods for Non-linear Reactive Flows

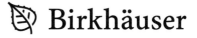 Birkhäuser

Murat Uzunca
Institute of Applied Mathematics
Middle East Technical University
Ankara, Turkey

Lecture Notes in Geosystems Mathematics and Computing
ISBN 978-3-319-30129-7   ISBN 978-3-319-30130-3   (eBook)
DOI 10.1007/978-3-319-30130-3

Library of Congress Control Number: 2016936686

Mathematics Subject Classification (2010): 65M60, 65M50, 65N30, 35K58

Printed on acid-free paper

This book is published under the trade name Birkhäuser.
The registered company is Springer International Publishing AG Switzerland (www.birkhauser-science.com)

To my family

Acknowledgments

First of all, I would like to give special thanks to my supervisor Prof. Dr. Bülent Karasözen for his valuable guidance, useful suggestions, his penetrating advice and his kindness and patience. He has provided me the opportunity to get inside and understand the world of Numerical Analysis and Applied Mathematics. I would not have been able to succeed in completing my PhD studies without his golden support.

I am thankful to Prof. Dr. Willi Freeden and Clemens Heine for their suggestions and kind opportunity for my book to be included in this series.

I acknowledge Dr. Hamdullah Yücel for discussions on discontinuous Galerkin methods.

I acknowledge Dr. Martin Weiser and Lars Lubkoll for their kind hosting in Berlin when we collaborated to study their own package program KASKADE.

I acknowledge Prof. Dr. Fredi Tröltzsch, Prof. Dr. Ronald W. Hoppe and Prof. Dr. Peter Deuflhard for their suggestions and valuable discussions.

I give special thanks to Assis. Prof. Dr. İsmail Uğur Tiryaki and Assis. Prof. Dr. Tahire Özen Öztürk for their guidance and suggestions which make me decide to start my PhD study at Middle East Technical University.

I would like to thank Ersin Kızgut, Ayşegül Kıvılcım, Adalet Çengel, Sibel Doğru Akgöl and all other Friends, all Members of the Mathematics Department and all Members of the Institute of Applied Mathematics at Middle East Technical University for the pleasant and friendly atmosphere they provided.

Finally, special thanks to my wife Mukaddes Kaya Uzunca and to my Family for their support and patience through all my educational life.

Contents

Chapter 1
Introduction

1.1 Geological and Computational Background

Production of drinkable water for human consumption and for other supply pur-
poses, as well as exploitation of available geo-resources by human intervention,
are the main characteristics of the geological subsurface. Protection and sustainable
management of water resources is one of the key problems in environmental engineering. Modeling and forecasting of soil and ground water contamination in industrial areas and in large scale agricultural land pose new challenges in geosciences. In
this aspect, accurate modeling and simulation of coupled ground and surface water
flows is a necessity. Chemically reactive components such as dissolved minerals,
colloids, or contaminants are transported by advection and diffusion over long dis-
tances through some highly heterogeneous porous media. Fundamental processes
in many geo-engineering applications, including oil and gas recovery, groundwater contamination and sustainable use of groundwater resources, storing greenhouse
gases (e.g. CO_2) or radioactive waste in the subsurface, are described by advection diffusion reaction (ADR) equations. Physical processes in various subdomains
of ground and hydrosystems differ. Therefore, different model concepts have to be
chosen for the subdomains which take into account the interaction between the subdomains, leading to coupled subsurface problems. Advanced numerical methods
are needed for accurate simulation of the coupled surface and subsurface water over
large space and time scales; inappropriate numerical methods can lead to false pre-
dictions.

Reactive transport modeling is characterized by material flow, transport, and reactions at multiple spatial and time scales, which is common in many branches
of geosciences. The transport, chemical, mechanical, and biological processes are
coupled in many cases; for example the advective and diffusive transport is com-
bined with local biogeochemical reactions providing electron donors and acceptors
[15, 16, 83]. The theoretical and numerical treatment of multi-component reactive
transport models date back to the mid-1980s, now encompassing multi-phase flow
and multicomponent reaction in fractured media [50], and in heterogeneous porous

© Springer International Publishing Switzerland 2016
M. Uzunca, *Adaptive Discontinuous Galerkin Methods for Non-linear Reactive Flows*,
Lecture Notes in Geosystems Mathematics and Computing,
DOI 10.1007/978-3-319-30130-3_1

media [79]. Groundwater contamination by biodegrading organic compounds is a serious and widespread environmental problem in many countries. Due to this contamination, the physical properties of the medium are changed through biological growth. Degradation of the contaminants is controlled to a large extent by biological and geochemical conditions in the groundwater, known as biodegradation. Whether passive remediation by natural processes of attenuation is sufficient or active remediation is needed can be decided using efficient, accurate and reliable numerical methods [15, 16]. Mineral precipitation or dissolution results in a feedback mechanism between flow, transport, and reaction. Reactive transport modeling and simulation is then an important tool for forecasting the complex interactions between coupled biological, chemical and geological processes and the effects of multiple space and time scales in the subsurface flow.

Numerical simulation of reactive transport in porous media is based on the models coupling of processes of different fields in the Earth sciences including hydrology, geochemistry, biogeochemistry, soil physics, and fluid dynamics. Reactive transport modeling nowadays covers many problems including multi-phase flow and multi-component flow in fractured and heterogeneous media. The permeability in heterogeneous porous and fractured media typically varies over orders of magnitude in space and time. Therefore the quantification and forecasting of these processes accurately is not an easy task. Highly variable flow is dominated entirely by either advection (Péclet number larger than one) or diffusion (Péclet number less than one) [85] . This causes macroscopic mixing and anomalous transport that is characterized by breakthrough of solutes or contaminants. Similarly chemical reaction rates can vary leading to complex mixing-induced reaction patterns, where the chemical reaction rates can dominate locally over transport rates . Therefore efficient and accurate numerical methods are needed to solve the ADR equations which resolve the wide range in flow velocities and reaction rates to predict spreading in space and mixing of reactive solutes. In addition, the exact spatial distribution of the permeability field and reaction rates are usually not known. All these features of the subsurface flow require a large number of simulations to quantify the uncertainty in fluid transport and to forecast the possible contamination of groundwater. The efficient modeling and simulation of flow in oil reservoirs with strongly varying heterogeneity, permeability and fine-scale fluctuations has a significant impact on reservoir performance prediction [49]. Additionally the efficiency of gas and oil injection is affected by mobility differences. Therefore an accurate solution of components in miscible gas injection processes occurring in the form of strongly nonlinear multiphase, multicomponent systems, have great economic and environmental relevance.

The ADR equations were discretized in space by all known methods such as finite differences, finite volumes, or finite elements, documented in a large body of literature. However, how to integrate in time the large stiff ordinary differential equations resulting from spatial discretization, and to solve problems with transport and reaction processes evolving over multiple time with sharp gradients in time and space, in a stable, accurate and efficient way by avoiding the non-physical oscillations remain as fundamental numerical challenges. There are many cases like reactive transport modeling and transport in heterogeneous media, where local small-scale features affect strongly the behavior of global solutions in large computational domains. If simple uniform meshes in space and time are used to achieve sufficient resolution in the many geoscience problems, even with ever increasing computational resources, the resolution of the discrete models can not be increased. Therefore different adaptive meshing strategies, including finite volume, finite element and discontinuous Galerkin (dG) methods (for a review see [28, 33]), in space and time have been developed in recent years in order to minimize the computational cost and to achieve high resolution in space and time.

In the last twenty years, dG methods have gained an increasing importance for an efficient and accurate solution of various kinds of partial differential equations. In contrast to the stabilized continuous Galerkin finite element methods, dG methods produce stable discretization without the need for extra stabilization strategies, and damp the unphysical oscillations for linear advection dominated problems. The dG combines the best properties of the finite volume and continuous finite elements methods. Finite volume methods can only use lower degree polynomials, and continuous finite elements methods require higher regularity due to the continuity requirements. Application of the adaptive dG methods and a posteriori error estimates to the problems in geosciences are reviewed recently for the multi-phase Darcy flow problem in [33]. Most applications of dG methods in geosciences concern reactive transport with advection [13, 62, 84], coupled surface-subsurface flow and transport problems [75], miscible displacement occurring in contaminated ground water and petroleum reservoirs [42], strong permeability contrasts such as layered reservoirs [89] and vanishing and varying diffusivity posing challenges in computations [72]. A different application of the space-time adaptive dG method in geosciences is presented in Castro et al. [28] for the computation of tsunami and seismic waves propagation.

1.2 Outline

We consider the following prototype ADR equation as a semi-linear PDE:

$$\frac{\partial u_i}{\partial t} - \varepsilon_i \Delta u_i + \beta_i \cdot \nabla u_i + r_i(\mathbf{u}) = f_i \qquad \text{in } \Omega_i \times (0,T], \qquad (1.1a)$$

$$u_i(x,t) = g_i^D \qquad \text{on } \Gamma_{D,i} \times (0,T], \qquad (1.1b)$$

$$\varepsilon_i \frac{\partial u_i}{\partial \mathbf{n}}(x,t) = g_i^N \qquad \text{on } \Gamma_{N,i} \times (0,T], \qquad (1.1c)$$

$$u_i(x,0) = u_i^0 \qquad \text{in } \Omega_i \qquad (1.1d)$$

for $i = 1, \ldots, m$, with Ω_i bounded, open, convex domains in $\mathbb{R}^2$ with boundaries $\partial \Omega_i = \Gamma_{D,i} \cup \Gamma_{N,i}, \Gamma_{D,i} \cap \Gamma_{N,i} = \emptyset, 0 < \varepsilon_i \ll 1$ are the diffusivity constants, $f_i \in L^2(\Omega)$ are the source functions, $\beta_i \in \left(W^{1,\infty}(\Omega)\right)^2$ are the velocity fields, $g_i^D \in H^{3/2}(\Gamma_{D,i})$ are the Dirichlet boundary conditions, $g_i^N \in H^{1/2}(\Gamma_{N,i})$ are the Neumann boundary conditions, $u_i^0 \in L^2(\Omega)$ are the initial conditions and $\mathbf{u}(x) = (u_1, \ldots, u_m)^T$ and $\mathbf{n}$ denote a vector of unknowns and an outward unit normal vector to the boundary, respectively.

The non-linear reaction terms $r_i(\mathbf{u})$ occur in chemical and environmental engineering, usually in the form of products and rational functions of concentrations, or exponential functions of the temperature, expressed by the Arrhenius law. Such models describe chemical and biological processes [15, 16] and they are strongly coupled because an inaccuracy in one unknown affects all the others. The applications in geoscience mostly concern reactive transport with advection [13, 62, 84] and strong permeability contrasts such as layered reservoirs [90] or vanishing and varying diffusivity posing challenges in computations [72]. The permeability in heterogeneous porous and fractured media varies over orders of magnitude in space, which results in a highly variable flow field, where the local transport is dominated by advection or diffusion [85]. Accurate and efficient numerical solutions of advection-diffusion-reaction equations predict the macroscopic mixing, anomalous transport of the solutes and contaminants for a wide range of parameters like permeability and Péclet numbers, different flow velocities and reaction rates are all challenging problems [85]. In order to resolve the complex flow patterns accurately, higher order time stepping methods like exponential time stepping methods are used (Tambue et al. 2010). We show here that, using space-time adaptivity, first-order backward Euler in time and dG in space, the same results can be obtained.

For the advection/reaction-dominated ADR equations, the standard Galerkin finite element methods are known to produce spurious oscillations, especially in the presence of sharp fronts in the solution, on boundary and interior layers. In the last two decades, several stabilization and shock/discontinuity capturing techniques were developed for linear and non-linear stationary and time dependent problems of type (1.1). In the linear advection-dominated case, the streamline upwind Petrov-Galerkin (SUPG) methods and dG methods are capable of handling the nonphysical oscillations due to the advection. Nevertheless, in the non-linear stationary case, the non-linear reaction term produces sharp layers in addition to the spurious oscillations due to the advection. An accurate and efficient numerical resolution of such layers is a challenge as the exact location of the layers are not known a priori. For the non-linear stationary problems, SUPG is used with the anisotropic shock capturing technique as SUPG-SC for reactive transport problems [15, 16, 17, 96]. It was shown that SUPG-SC is capable of reducing the unphysical oscillations in crosswind direction. However, the parameters of the SUPG and SUPG-SC should be designed carefully for efficient solution of the discretized equations. A comprehensive review of weighted residual methods, the orthogonal collocation, Galerkin, tau and least squares methods are given in [82] for solving linear and non-linear Pellet equations, where the methods are compared with respect to convergence of residuals and computational efficiency. On the other hand, in the non-stationary case, the resolution of spatial layers is more critical since the nature of sharp layers may vary as time progresses, and it is highly possible that there occur also temporal layers in addition to the spatial one.

In contrast to stabilized continuous Galerkin finite element methods, dG methods produce stable discretizations without the need for extra stabilization strategies, and damp unphysical oscillations for linear advection dominated problems. The dG combines the best properties of finite volume and continuous finite elements methods. Finite volume methods can only use lower degree polynomials, and continuous finite elements methods require higher regularity due to continuity requirements. In [98], several non-linear steady-state advection dominated problems are solved with dG-SC (dG method with the shock-capturing technique) in [70]. Therein, construction of shock-capturing parameters are problem dependent and require solution of the system several times, as in the case of SUPG-SC, and even then there still occur some unphysical oscillations. The dG method is in particular suitable for non-matching grids and hp (space and order) adaptivity [56], detecting sharp layers and singularities. They are easily adapted locally for non-conforming finite elements requiring less regularity. Higher order dG approximation can be easily used by hierarchical bases [25], and the convergence rates are limited by the consistency error which makes the dG suitable for complex fluid flows. The dG methods are robust with respect to variation of physical parameters like diffusion constant and permeability. The stability of the dG approximation retained by adjusting the penalty parameter to penalize jumps at the interface of the elements. A unified analysis of the interior penalty dG methods for elliptic PDEs are given in [5]. Other advantages of the dG methods are conservation of mass and fluxes and parallelization. Moreover, dG methods are better suited for adaptive strategies which are based on a posteriori error estimation.

For an accurate and efficient solution of the advection/reaction dominated non-stationary semi-linear ADR equations (1.1), we use a dG method for space discretization and backward Euler as the time integrator. We develop adaptive algorithms based on residual-based a posteriori error estimation both in space and time to save both spatial degrees of freedom (DoFs) and the number of time-steps. Using a numerical solution, a posteriori error bounds are derived, then the regions where the error is too large are located and refined adaptively. For the steady-state linear problems, there are a variety of well-understood a posteriori error estimation studies on pure diffusion problems using standard finite elements methods (FEMs) [3, 91] and dG methods [18, 23, 55, 60]. For the steady-state linear advection-diffusion problems we refer to [43, 56, 62, 63, 33, 78, 80, 93, 92] and references therein. In case of the non-stationary models, there are several studies for linear diffusion models [44, 47] and linear advection-diffusion equations [37] using dG discretization in space. There exist also some space-time adaptive methods with interior penalty Galerkin (IPG) and efficient time integrators for linear advection-diffusion equations [36, 35, 38].

It is well-known that energy techniques to derive a posteriori estimates in non-stationary problems are challenging. First of all, the discrete residual usually does not make sense, and even leads to singular right-hand sides. Moreover, it is known that the derived a posteriori estimates are optimal in $L^2(H^1)$-type norms but suboptimal in $L^\infty(L^2)$-type norms. Here, we utilize an elliptic reconstruction technique [67] by which we derive a posteriori estimates by energy techniques for non-stationary models of the form (1.1) using the a posteriori error estimates derived for the stationary (elliptic) case of (1.1), which, to the best of our knowledge, has not been applied yet. The idea of the elliptic reconstruction technique is to construct an auxiliary solution whose difference between the numerical solution can be estimated by a known (elliptic) a posteriori estimate, and the constructed auxiliary solution satisfies a variant of the given problem with a right-hand side which can be controlled in an optimal way. In this way, in contrast to the standard energy techniques, we do not need to try to adapt the estimates case by case in order to compare the exact solution with the numerical solution directly, and we are able to obtain results being optimal order in both $L^2(H^1)$ and $L^\infty(L^2)$-type norms, while the results obtained by the standard energy methods are only optimal order in $L^2(H^1)$-type norms, but sub-optimal order in $L^\infty(L^2)$-type norms.

The rest of the book is organized as follows. In Chapter 2 we give a detailed description of the interior penalty discontinuous Galerkin (IPG) methods. Since IPG schemes concerns with the diffusion part of the problem, we start with the Poisson problem. After discussion of the effect of penalty parameter on spatial accuracy, in Chapter 2, we give the IPG formulation for a general steady-state linear ADR equation with upwinding for advection, which forms the key part of the non-linear elliptic model to be introduced in Chapter 3.

In Chapter 3, we introduce the steady-state (elliptic) form of the non-stationary model (1.1). We give the existence and uniqueness results for stationary semi-linear ADR equations using coercivity results for the bilinear forms in Chapter 2 and the assumptions on the non-linear reaction term. Then, we describe the adaptive algorithm for semi-linear ADR equations. We derive robust (in the diffusion parameter or in Péclet number) a posteriori error bounds. To derive the a posteriori error bounds, we utilize the robust a posteriori error estimators in [80] for steady-state linear advection-diffusion equations. Because the condition number of the stiffness matrices grows rapidly with the number of elements and with the penalty parameter for dG discretized equations, efficient solution strategies such as preconditioning are required to solve the resulting linear systems. We introduce, in Chapter 3, an iterative method called matrix reordering [88] which is robust and efficient. In numerical studies, we demonstrate the efficiency of the matrix reordering iterative method in detail by comparing the CPU times, number of iteration numbers etc. with the ones obtained by the direct solvers. Further, in Chapter 3, we will compare the adaptive dG approximations with the approximations by a famous stabilized FEM, Galerkin least squares FEM [22, 21, 58] which also can be compared with the method SUPG. Chapter 3 ends with some numerical studies on the stationary semi-linear models by which we demonstrate that the adaptive dG schemes are capable of resolving the solution of the elliptic problems at the layers, and they are more accurate compared to other methods such as SUPG-SC [15, 16] and dG-SC [98].

In Chapter 4, we introduce the dG discretized semi-discrete formulation of the model (1.1), and we give the existence and uniqueness results. Then, we introduce the elliptic reconstruction technique [67] by comparing with the usual energy techniques. Using the elliptic reconstruction technique, we state the a posteriori error bounds for the semi-discrete system using the a posteriori error bounds driven and proven in Chapter 3 for stationary problems. After giving an in time backward Euler discretized fully-discrete formulation of the model problem (1.1), we introduce the a posteriori error bounds for in space SIPG and in the time backward Euler discretized fully-discrete system of (1.1) utilizing the related bounds for a semi-discrete system using $L^{\infty}(L^2) + L^2(H^1)$-type norm, and we give a modification of robust adaptive algorithm both in space and time [26] to the non-stationary semi-linear problems of type (1.1). We also state the solution of linear systems arising from Newton's method algebraically. Chapter 4 follows with numerical studies demonstrating effectiveness of the adaptive algorithm. We show that our adaptive algorithm is robust in a diffusion parameter, similar to the linear non-stationary ADR equations given in [26], by demonstrating both the spatial and temporal effectivity indices and convergence rates for a wide range of diffusion parameters including the advection domination cases. We also point out that our adaptive algorithm is capable of catching not only the spatial layers but also the temporal layers by decreasing the time-step size there. In addition, by some examples of flow transport, we determine that beside the refinement procedure, the coarsening procedure of the adaptive algorithm works effectively, which is crucial to save time and DoFs. In Appendix A, you can find a MATLAB routine to solve the semi-linear ADR equations using dG methods to discretize in space.

Chapter 2
Discontinuous Galerkin Methods

The discontinuous Galerkin (dG) method was introduced by Reed and Hill [73] in 1973 for steady-state neutron transport as an hyperbolic problem. This was followed by other studies; by Bassi and Rebay [12] for the compressible Navier-Stokes equations, Cockburn and Shu [31] developed the local discontinuous Galerkin (ldG) method for advection-diffusion equations, and Peraire and Persson [69] introduced the compact discontinuous Galerkin (cdG) method. Independent of the dG methods, interior penalty (IP) methods have been developed for elliptic and parabolic problems by Douglas and Dupont [40] and Wheeler [95]. Then, in the 1980's, Arnold et al. [6] proposed a unified classification and analysis of various kinds of dG methods. Later on, the dG methods were developed for elliptic problems [8, 24, 77] and for problems with advection [7, 14, 51, 56].

In the last two decades, dG methods have become so popular as an alternative to the finite volume method (FVM) and the continuous finite element (FEM) method. FVMs can only use lower degree polynomials, and continuous FEMs require higher regularity due to the continuity requirements. The dG methods combine the best properties of FVMs and continuous FEMs such as consistency, flexibility, stability, conservation of local quantities, robustness and compactness. The consistency of dG methods can be easily interpreted using the Galerkin orthogonality property of FEMs. The flexibility of dG methods comes from the fact that the functions in space are discontinuous along the inter-element boundaries, which is a key element in the generation of unstructured meshes with hanging nodes and on the construction of higher order basis functions. dG discretization allows us to easily use different orders of polynomials on different elements since the supports of the functions in space are just a single element and there is no overlapping between the elements as in classical FEMs. Therefore, dG methods can be easily used in p-refinement schemes, where the order of the polynomials on the elements having steep gradient can be arranged adaptively. The stability of dG methods are handled via the penalty term which penalizes the jumps of the solution on the element boundaries. In this way, the stability in dG methods are inherited and do not require additional stabilization techniques as in the streamlined upwind Petrov-Galerkin (SUPG) method being the most popular continuous FEM method for advection dominated problems.

© Springer International Publishing Switzerland 2016
M. Uzunca, *Adaptive Discontinuous Galerkin Methods for Non-linear Reactive Flows*,
Lecture Notes in Geosystems Mathematics and Computing,
DOI 10.1007/978-3-319-30130-3_2

Due to local structure of dG methods, physical quantities such as mass and energy are conserved locally through dG schemes, which is important for flow and transport problems. Sharp layers and singularities can be detected locally via the fully discontinuous polynomial representation of the solution, which makes dG methods extremely convenient for adaptive *h*- (space) refinement and also *hp*-(space and order) refinement [81]. The dG methods are designed to be robust for perturbation of the parameters such as diffusion and advection constants. The perturbation of the parameters affect the solution only locally and it will remained unperturbed in the remaining field. In addition, the Dirichlet boundary conditions are imposed weakly in contrast to the continuous FEMs, where additional conditions have to be imposed on the boundary. All these properties make the dG methods in a compact form as an efficient and accurate discretization technique for parallel computing. Despite all mentioned advantages, the dG methods have some drawbacks. Compared to the continuous FEMs, dG discretization produces dense and ill-conditioned matrices increasing with the order of polynomial degree.

   In this chapter, we give a detailed description of the interior penalty discontinuous Galerkin (IPG) methods. As a model problem, we consider the general Poisson problem in Section 2.2. In Section 2.3, the computational tools are introduced for the dG discretization such as finite element spaces, basis function etc. The effect of the penalty parameter on the accuracy of solutions is discussed in Section 2.4. At the end of the chapter, we give an arising scheme for a general steady-state linear ADR equation in Section 2.5, which forms the stationary linear part of the scalar form of the model problem (1.1) used in this book.

2.1 Preliminaries

In this section, we introduce some useful definitions and identities required in the construction of IPG schemes and to show the coercivity of the bilinear form arising from it.

2.1.1 Sobolev Spaces

On a polygonal domain Ω in $\mathbb{R}^d$, the spaces $L^p(\Omega)$ of p-integrable functions are defined by

$$L^p(\Omega) = \{v \text{ Lebesgue measurable } : \ \|v\|_{L^p(\Omega)}^2 < \infty\}, \qquad 1 \leq p \leq \infty$$

equipped with the norms

$$\|v\|_{L^p(\Omega)} = \left(\int_\Omega |v(x)|^p dx \right)^{1/p}, \qquad\qquad 1 \le p < \infty,$$

$$\|v\|_{L^\infty(\Omega)} = esssup\{|v(x)| : x \in \Omega\}, \qquad\qquad p = \infty.$$

We mainly consider the space $L^2(\Omega)$ which is a Hilbert space equipped with the usual L^2-inner product

$$(u,v)_\Omega = \int_\Omega u(x)v(x)dx \, , \ \|v\|_{L^2(\Omega)} = \sqrt{(v,v)_\Omega} .$$

Let $\mathscr{D}(\Omega)$ denote the subspace of the space C^∞ having compact support in Ω. For any multi-index $\alpha = (\alpha_1, \ldots, \alpha_d) \in \mathbb{N}^d$ with $|\alpha| = \sum_{i=1}^d \alpha_i$, the distributional derivative $D^\alpha v$ is defined by

$$D^\alpha v(\psi) = (-1)^{|\alpha|} \int_\Omega v(x) \frac{\partial^{|\alpha|} \psi}{\partial x_1^{\alpha_1} \cdots \partial x_d^{\alpha_d}} \, , \quad \forall \psi \in \mathscr{D}(\Omega).$$

Then, we introduce for an integer s the Sobolev spaces

$$H^s(\Omega) = \{v \in L^2(\Omega) : D^\alpha v \in L^2(\Omega) \, , \ \forall 0 \le |\alpha| \le s\}$$

with the associated Sobolev norm

$$\|v\|_{H^s(\Omega)} = \left(\sum_{0 \le |\alpha| \le s} \|D^\alpha v\|_{L^2(\Omega)}^2 \right)^{1/2},$$

and the associated Sobolev seminorm

$$|v|_{H^s(\Omega)} = \|\nabla^s v\|_{L^2(\Omega)} = \left(\sum_{|\alpha|=s} \|D^\alpha v\|_{L^2(\Omega)}^2 \right)^{1/2}.$$

The Sobolev spaces with vanishing functions on the domain boundary are defined by

$$H_0^s(\Omega) = \{v \in H^s(\Omega) : v|_{\partial\Omega} = 0\}.$$

We are mainly interested in the case $s = 1$ where

$$H^1(\Omega) = \{v \in L^2(\Omega) : \nabla v \in (L^2(\Omega))^d\}.$$

Moreover, for a partition (for example triangles) ξ_h of Ω, the broken Sobolev spaces are given by

$$H^s(\xi_h) = \{v \in L^2(\Omega) : v|_K \in H^s(K) \, , \ \forall K \in \xi_h\}$$

with the associated broken Sobolev norm

$$\|v\|_{H^s(\xi_h)} = \left(\sum_{K \in \xi_h} \|v\|^2_{H^s(K)} \right)^{1/2},$$

and the associated broken gradient seminorm

$$|v|_{H^0(\xi_h)} = \left(\sum_{K \in \xi_h} \|\nabla v\|^2_{L^2(K)} \right)^{1/2}.$$

2.1.2 Trace Theorems

Theorem 2.1. *[74, Theorem 2.5] There exist trace operators $\gamma_0: H^s(\Omega) \mapsto H^{s-1/2}(\partial \Omega)$ ($s > 1/2$) and $\gamma_1: H^s(\Omega) \mapsto H^{s-3/2}(\partial \Omega)$ ($s > 3/2$) being extensions of the boundary values and boundary normal derivatives, respectively. For the polygonal boundary $\partial \Omega$, and for $v \in C^1(\overline{\Omega})$, we have*

$$\gamma_0 v = v|_{\partial \Omega} , \ \gamma_1 v = \nabla v \cdot \mathbf{n}|_{\partial \Omega}.$$

In the above theorem, $H^{s-1/2}(\partial \Omega)$ ($H^{s-3/2}(\partial \Omega)$) is the space of completion of all functions in $H^s(\partial \Omega)$ ($H^{s-1}(\partial \Omega)$) with the property

$$H^s(\partial \Omega) \subset H^{s-1/2}(\partial \Omega) \subset H^{s-1}(\partial \Omega).$$

For instance, when $s = 2$, γ_0 (γ_1) belongs to the space $H^{3/2}(\partial \Omega)$ ($H^{1/2}(\partial \Omega)$), the interpolated space between the spaces $H^2(\partial \Omega)$ ($H^1(\partial \Omega)$) and $H^1(\partial \Omega)$ ($L^2(\partial \Omega)$). As a consequence of the above theorem, the trace inequalities are given as

$$\|v\|_{L^2(e)} \leq C_{Tr} h_K^{-1/2} \|v\|_{L^2(K)}, \tag{2.1a}$$

$$\|\nabla v \cdot \mathbf{n}\|_{L^2(e)} \leq C_{Tr} h_K^{-1/2} \|\nabla v\|_{L^2(K)}, \tag{2.1b}$$

where h_K denotes the diameter of an element K, and the positive constant C_{Tr} is independent of h_K.

2.1.3 Cauchy-Schwarz's and Young's Inequalities

The following inequalities are frequently used in the analysis of FEMs.

- Cauchy-Schwarz's inequality: For any $u, v \in L^2(\Omega)$,

$$|(u, v)_{L^2(\Omega)}| \leq \|u\|_{L^2(\Omega)} \|v\|_{L^2(\Omega)}. \tag{2.2}$$

- Young's inequality: For any $\delta > 0$ and for any $a, b \in \mathbb{R}$,

$$ab \leq \frac{\delta}{2}a^2 + \frac{1}{2\delta}b^2. \tag{2.3}$$

2.2 Construction of IPG Methods

In this section, we give a detailed description of the IPG methods [5, 74] for the Poisson equation

$$-\varepsilon \Delta u = f \qquad \text{in } \Omega \subset \mathbb{R}^2, \tag{2.4a}$$

$$u = g^D \qquad \text{on } \Gamma_D, \tag{2.4b}$$

$$\varepsilon \nabla u \cdot \mathbf{n} = g^N \qquad \text{on } \Gamma_N, \tag{2.4c}$$

with $\partial \Omega = \Gamma_D \cup \Gamma_N$ and $\Gamma_D \cap \Gamma_N = \emptyset$.

Let the mesh $\xi_h = \{K\}$ be a family of shape regular elements (triangles), i.e. for some positive constant h_0 there holds

$$\max_{K \in \xi_h} \frac{h_K^2}{|K|} \leq h_0,$$

where h_K and $|K|$ denote the diameter and the area of the element K, respectively and $\overline{\Omega} = \cup \overline{K}$ and $K_i \cap K_j = \emptyset$ for $K_i, K_j \in \xi_h$. We denote by Γ_h^0, Γ_h^D and Γ_h^N the set of interior, Dirichlet boundary and Neumann boundary edges, respectively, so that $\Gamma_h^0 \cup \Gamma_h^D \cup \Gamma_h^N$ forms the skeleton of the mesh. For any $K \in \xi_h$, let $\mathbb{P}_k(K)$ be the set of all polynomials of degree at most k on K. Then, let the finite dimensional solution and test function spaces be given by

$$V_h = \{v \in L^2(\Omega) : v|_K \in \mathbb{P}_k(K), \forall K \in \xi_h\} \not\subset H_0^1(\Omega).$$

We note that the trial and test function spaces are the same because the boundary conditions in dG methods are weakly imposed. The classical (continuous) FEM uses a conforming, finite-dimensional subspace $V_h \subset H_0^1(\Omega)$, which requires that the space V_h contains smooth functions (e.g., $V_h \subset \{v \in C(\overline{\Omega}) : v = 0 \text{ on } \partial\Omega\}$). On the other hand, dG methods use non-conforming spaces, in which case the functions in $V_h \not\subset H_0^1(\Omega)$ are allowed to be discontinuous on the inter-element boundaries.

Since the functions in V_h are discontinuous along the inter-element boundaries, along an interior edge, there are two different traces from the adjacent elements sharing that edge. We give first some notations before starting the construction of IPG formulation. Let $K_i, K_j \in \xi_h$ $(i < j)$ be two adjacent elements sharing an interior edge $e = K_i \cap K_j \subset \Gamma_h^0$ (see Fig.2.1). The trace of a scalar function v is denoted from inside K_i by v_i and from inside K_j by v_j. Then, we can set the jump and average values of v on the edge e,

$$[v] = v_i \mathbf{n}_e - v_j \mathbf{n}_e, \quad \{v\} = \frac{1}{2}(v_i + v_j),$$

where $\mathbf{n}_e$ is the unit normal to the edge e oriented from K_i to K_j. Similarly, we set the jump and average values of a vector-valued function $\mathbf{q}$ on e,

$$[\mathbf{q}] = \mathbf{q}_i \cdot \mathbf{n}_e - \mathbf{q}_j \cdot \mathbf{n}_e, \quad \{\mathbf{q}\} = \frac{1}{2}(\mathbf{q}_i + \mathbf{q}_j).$$

Observe that $[v]$ is a vector for a scalar function v, while, $[\mathbf{q}]$ is scalar for a vector valued function $\mathbf{q}$. On the other hand, along any boundary edge $e = K_i \cap \partial\Omega$, we set

$$[v] = v_i\mathbf{n}, \quad \{v\} = v_i, \quad [\mathbf{q}] = \mathbf{q}_i \cdot \mathbf{n}, \quad \{\mathbf{q}\} = \mathbf{q}_i,$$

where $\mathbf{n}$ is the unit outward normal to the boundary at e.

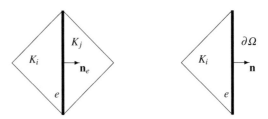

Fig. 2.1: Two adjacent elements sharing an edge (left); an element near to domain boundary (right)

Now, we can construct the IPG discretization of the Poisson equation by multi-plying (2.4a) by a test function $v \in V_h$, and integrating over Ω,

$$-\sum_{K \in \xi_h} \int_K \varepsilon \Delta u v\, dx = \sum_{K \in \xi_h} \int_K f v\, dx.$$

Applying the divergence theorem on every element integral gives

$$\sum_{K \in \xi_h} \int_K \varepsilon \nabla u \cdot \nabla v\, dx - \sum_{K \in \xi_h} \int_{\partial K} \varepsilon (\nabla u \cdot \mathbf{n}) v\, ds = \sum_{K \in \xi_h} \int_K f v\, dx + \sum_{e \in \Gamma_h^N} \int_e g^N v\, ds.$$

Alternatively, using the jump definitions ($v \in V_h$ are element-wise discontinuous), we obtain

$$\sum_{K \in \xi_h} \int_K \varepsilon \nabla u \cdot \nabla v\, dx - \sum_{e \in \Gamma_h^0 \cup \Gamma_h^D} \int_e [\varepsilon v \nabla u]\, ds = \sum_{K \in \xi_h} \int_K f v\, dx + \sum_{e \in \Gamma_h^N} \int_e g^N v\, ds.$$

One can easily verify that $[\varepsilon v \nabla u] = \{\varepsilon \nabla u\} \cdot [v] + [\varepsilon \nabla u] \cdot \{v\}$. Then, for $[\nabla u] = 0$, and assuming u is to be smooth enough so that ∇u is continuous), we obtain

$$\sum_{K\in\xi_h}\int_K \varepsilon\nabla u\cdot\nabla v\,dx - \sum_{e\in\Gamma_h^0\cup\Gamma_h^D}\int_e \{\varepsilon\nabla u\}\cdot[v]\,ds = \sum_{K\in\xi_h}\int_K fv\,dx + \sum_{e\in\Gamma_h^N}\int_e g^N v\,ds.$$

Yet, the left-hand side may not be coercive. To overcome this, the solutions on the inter-element boundaries are penalized. For $[u]=0$ along the interior edges and for continuous u, we obtain

$$\sum_{K\in\xi_h}\int_K \varepsilon\nabla u\cdot\nabla v\,dx - \sum_{e\in\Gamma_h^0\cup\Gamma_h^D}\int_e \{\varepsilon\nabla u\}\cdot[v]\,ds + \kappa\sum_{e\in\Gamma_h^0}\int_e \{\varepsilon\nabla v\}\cdot[u]\,ds$$

$$+ \sum_{e\in\Gamma_h^0}\frac{\sigma\varepsilon}{h_e}\int_e [u]\cdot[v]\,ds = \sum_{K\in\xi_h}\int_K fv\,dx + \sum_{e\in\Gamma_h^N}\int_e g^N v\,ds,$$

where h_e denote the length of the edge e and σ is the penalty parameter. Finally, we add to both sides the edge integrals on the Dirichlet boundary edges by keeping unknown on the left-hand side and imposing a Dirichlet boundary condition on the right-hand side

$$\sum_{K\in\xi_h}\int_K \varepsilon\nabla u\cdot\nabla v\,dx - \sum_{e\in\Gamma_h^0\cup\Gamma_h^D}\int_e \{\varepsilon\nabla u\}\cdot[v]\,ds + \kappa\sum_{e\in\Gamma_h^0\cup\Gamma_h^D}\int_e \{\varepsilon\nabla v\}\cdot[u]\,ds$$

$$+ \sum_{e\in\Gamma_h^0\cup\Gamma_h^D}\frac{\sigma\varepsilon}{h_e}\int_e [u]\cdot[v]\,ds = \sum_{K\in\xi_h}\int_K fv\,dx \tag{2.5}$$

$$+ \sum_{e\in\Gamma_h^D}\int_e g^D\left(\frac{\sigma\varepsilon}{h_e}v + \kappa\varepsilon\nabla v\cdot\mathbf{n}\right)ds + \sum_{e\in\Gamma_h^N}\int_e g^N v\,ds,$$

which gives the IPG formulation. The parameter κ in the IPG formulation (2.5) determines the type of the IPG method. For different values $\kappa \in \{-1,0,1\}$, the IPG methods are classified as:

$\kappa = -1$: Symmetric interior penalty Galerkin (SIPG) method,

$\kappa = 1$: Non-symmetric interior penalty Galerkin (NIPG) method,

$\kappa = 0$: Incomplete interior penalty Galerkin (IIPG) method.

In this book, we consider only the symmetric interior penalty Galerkin (SIPG) method by setting $\kappa = -1$.

Proposition 2.1. *The equivalence (consistency) of the model problem (2.4) and IPG variational problem (2.5) is obvious by the construction of the IPG formulation above. Under the assumption that the solution u of the model problem (2.4) satisfies $u \in H^s(\Omega)$ for some $s > 3/2$, discrete solutions constructed by the IPG method exist.*

2.3 Computation of the Integral Terms

In this section, we introduce some tools that can be used to compute integrals on physical elements. Firstly, we mention the *reference element* approach which is a common tool for any application of the finite elements method. Then, we discuss the basis functions used in IPG methods, and related numerical quadrature rules.

2.3.1 Reference Element

It is well-known that computing integrals on physical elements is difficult and costly. The common technique in FEMs, instead, is to compute all the integrals on a reference element and move them to the physical elements.

We use the unit triangle on the first quadrant as the reference triangle $\hat{K}$ with vertices $\hat{A}_1 = (0,0)$, $\hat{A}_2 = (1,0)$, $\hat{A}_3 = (0,1)$, while a physical element K has the vertices $A_i(x_i, y_i)$ for $i = 1,2,3$. To compute the integrals on the physical elements, we utilize the invertible affine mapping $F_K : \hat{K} \mapsto K$ defined by

$$F_K \begin{pmatrix} \hat{x} \\ \hat{y} \end{pmatrix} = \begin{pmatrix} x \\ y \end{pmatrix}, \qquad x = \sum_{i=1}^{3} x_i \hat{\psi}_i(\hat{x}, \hat{y}), \; y = \sum_{i=1}^{3} y_i \hat{\psi}_i(\hat{x}, \hat{y}),$$

with the shape functions

$$\hat{\psi}_1(\hat{x}, \hat{y}) = 1 - \hat{x} - \hat{y}, \; \hat{\psi}_2(\hat{x}, \hat{y}) = \hat{x}, \; \hat{\psi}_3(\hat{x}, \hat{y}) = \hat{y}.$$

The mapping can be written as

$$\begin{pmatrix} x \\ y \end{pmatrix} = F_K \begin{pmatrix} \hat{x} \\ \hat{y} \end{pmatrix} = B_K \begin{pmatrix} \hat{x} \\ \hat{y} \end{pmatrix} + b_K,$$

where B_K is a non-singular matrix and b_K is a translation vector given by

$$B_K = \begin{pmatrix} a_{11}^K & a_{12}^K \\ a_{21}^K & a_{22}^K \end{pmatrix} = \begin{pmatrix} x_2 - x_1 & x_3 - x_1 \\ y_2 - y_1 & y_3 - y_1 \end{pmatrix}, \qquad b_K = \begin{pmatrix} x_1 \\ y_1 \end{pmatrix}.$$

Thus, the inverse of the affine map F_K is defined explicitly as

$$F_K^{-1} : K \mapsto \hat{K} : \qquad F_K^{-1}(x) = B_K^{-1}(x - b_K) = \hat{x},$$

where the inverse matrix B_K^{-1} is given by

$$B_K^{-1} = \frac{1}{\det B_K} \begin{pmatrix} a_{22}^K & -a_{12}^K \\ -a_{21}^K & a_{11}^K \end{pmatrix} = \frac{1}{2|K|} \begin{pmatrix} \hat{a}_{11}^K & \hat{a}_{12}^K \\ \hat{a}_{21}^K & \hat{a}_{22}^K \end{pmatrix} = \frac{1}{2|K|} \hat{B}_K.$$

Using the definitions above we obtain the identities between the functions on the reference element and the functions on the physical elements

$$\hat{v}(\hat{x},\hat{y}) = v(x,y),$$
$$\hat{\nabla}\hat{v}(\hat{x},\hat{y}) = B_K^T \nabla v(x,y).$$

2.3.2 Numerical Quadrature

The flexibility of dG methods allow us to use high order polynomials. High order quadrature rules should be used to compute the integrals accurately, while the explicit integral formulation is complicated with high order polynomials. We use the numerical quadrature rule in [41] to approximate the integrals on the reference element $\hat{K}$

$$\int_{\hat{K}} \hat{v} \approx \sum_{j=1}^{N_q} w_j \hat{v}(s_{x,j}, s_{y,j}),$$

where w_j's denote the quadrature weights and $(s_{x,j}, s_{y,j}) \in \hat{K}$ are the quadrature nodes inside the reference element. Use of the affine map F_K with the quadrature formula above leads to computation of the integrals on a physical element K as

$$\int_K v = \int_{\hat{K}} v \circ F_K \det(B_K) = 2|K| \int_{\hat{K}} \hat{v} \approx 2|K| \sum_{j=1}^{N_q} w_j \hat{v}(s_{x,j}, s_{y,j}),$$

$$\int_K \nabla v \cdot w \approx 2|K| \sum_{j=1}^{N_q} w_j (B_K^T)^{-1} \hat{\nabla}\hat{v}(s_{x,j}, s_{y,j}) \cdot \hat{w}(s_{x,j}, s_{y,j}),$$

$$\int_K \nabla v \cdot \nabla w \approx 2|K| \sum_{j=1}^{N_q} w_j (B_K^T)^{-1} \hat{\nabla}\hat{v}(s_{x,j}, s_{y,j}) \cdot (B_K^T)^{-1} \hat{\nabla}\hat{w}(s_{x,j}, s_{y,j}).$$

2.3.3 Basis Functions

The space of functions of the dG solutions is given as

$$V_h = \text{span}\{\psi_i^K : 1 \leq i \leq N_{loc}, \ K \in \xi_h\}$$

with the global basis functions

$$\psi_i^K(x) = \begin{cases} \hat{\psi}_i \circ F_K(x), & \text{if } x \in K, \\ 0, & \text{if } x \notin K, \end{cases}$$

where $\{\hat{\psi}_i\}$'s are the local basis functions defined on the reference element $\hat{K}$ and N_{loc} denotes the local dimension depending on the order k of the polynomial basis functions, and it is given by $N_{loc} = (k+1)(k+2)/2$ for a 2D spatial domain.

There are a variety of basis functions such as Lagrange shape functions, monomial bases, Legendre polynomials etc. In this book, we use the orthogonal Dubiner basis [32] defined on the reference triangle

$$\hat{K} = \{x = (x_1, x_2) | \ 0 \le x_1, x_2 \le 1\}.$$

The construction of such basis polynomials are based on the collapsed coordinate transform between the reference triangle $\hat{K}$ and the reference square $\hat{Q} = [-1, 1]^2$ (see Fig.2.2). First, the basis polynomials on the square $\hat{Q}$ are formed by a general-

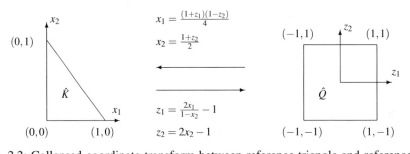

Fig. 2.2: Collapsed coordinate transform between reference triangle and reference square

ized tensor product of the Jacobi polynomials on the interval $[-1, 1]$, and then, these basis polynomials are transformed to the reference triangle $\hat{K}$ using the collapsed coordinate transform in Fig.2.2. The explicit form of Dubiner basis polynomials on the reference triangle $\hat{K}$ is given by

$$\phi_{mn}(x_1, x_2) = (1 - z_2)^m P_m^{0,0}(z_1) P_n^{2m+1,0}(z_2), \qquad 0 \le m, n, m+n \le Nloc,$$

$$= 2^m (1 - x_2)^m P_m^{0,0}\left(\frac{2x_1}{1 - x_2} - 1\right) P_n^{2m+1,0}(2x_2 - 1),$$

where $P_n^{\alpha,\beta}(x)$'s denote the corresponding n-th order Jacobi polynomials on the interval $[-1, 1]$, which, under the Jacobi weight $(1 - x)^\alpha (1 + x)^\beta$, are orthogonal polynomials, i.e.,

$$\int_{-1}^{1} (1 - x)^\alpha (1 + x)^\beta P_m^{\alpha,\beta}(x) P_n^{\alpha,\beta}(x) dx = \delta_{mn}.$$

This property of the Jacobi polynomials yields the orthogonality of the Dubiner basis on the reference triangle $\hat{T}$ as

$$\iint_{\hat{T}} \phi_{mn}(x_1, x_2) \phi_{ij}(x_1, x_2) dx_1 dx_2 = \frac{1}{8} \delta_{mi} \delta_{nj}.$$

The orthogonality of the Dubiner basis leads to diagonal mass matrix and a better-conditioned stiffness matrix compared to the other basis polynomials (see Fig.2.3), and for higher order polynomial basis, the integrals are approximated with high accuracy.

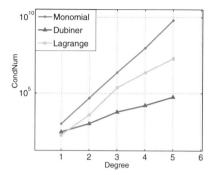

Fig. 2.3: Degree vs. condition number of the stiffness matrix: comparison for different types of basis functions for the Poisson problem (2.4)

2.4 Effect of Penalty Parameter

The penalty parameter σ in the SIPG formulation should be selected sufficiently large to ensure the coercivity of the bilinear form [74, Sec. 27.1], which is needed for the stability of the convergence of the SIPG method. It ensures that the matrix arising from the SIPG discretization of the diffusion part is symmetric positive definite. At the same time it should not be too large since the condition number of the stiffness matrices obtained by discretizing the bilinear form increases linearly with the penalty parameter (see Fig.2.4, left). In the literature, several choices of the penalty parameter are suggested. In [42], computable lower bounds are derived, and in [34], the penalty parameter is chosen depending on the diffusion coefficient ε. The effect of the penalty parameter on the condition number was discussed in detail for the dG discretization of the Poisson equation in [27] and in [90] for layered reservoirs with strong permeability contrasts, e.g. ε varying between 10^{-1} and 10^{-7}.

To examine the effect of the penalty parameter, we study the Poisson problem (2.4a) with the appropriate load function f, diffusion constant $\varepsilon = 1$ and Dirichlet boundary conditions using the exact solution $u(x) = \sin(\pi x_1)\sin(\pi x_2)$. In Fig.2.5, we have plotted the maximum nodal errors depending on the penalty parameter to show the instability of the scheme for different degrees of bases, where the triangular symbols indicate our choice $\sigma = 3k(k+1)$.

Similarly, the condition number of the stiffness matrix increases with decreasing mesh-size and with increasing order of the dG discretization (see Fig.2.4, right),

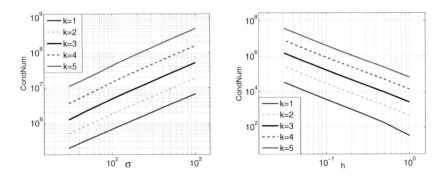

Fig. 2.4: Condition number of the stiffness matrix of the SIPG method as the functions of the penalty parameter σ (left) and the mesh-size h (right) with different polynomial degree k for the Poisson problem (2.4)

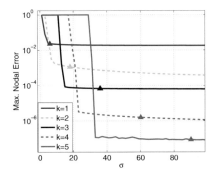

Fig. 2.5: Maximum nodal errors of the SIPG approximation as a function of penalty parameter σ with different polynomial degree k for the Poisson problem (2.4)

which affects the efficiency of an iterative solver. Similar results can be found also in [27].

2.5 dG Discretization of Advective Terms

In this Section, we consider the scalar advection-diffusion equation of the form

$$-\varepsilon \Delta u + \beta \cdot \nabla u + \alpha u = f \qquad \text{in } \Omega, \tag{2.6a}$$

$$u = g^D \qquad \text{on } \Gamma_D, \tag{2.6b}$$

$$\varepsilon \frac{\partial u}{\partial \mathbf{n}} = g^N \qquad \text{on } \Gamma_N \tag{2.6c}$$

where Ω is a bounded, open, convex domain in $\mathbb{R}^2$ with boundary $\partial \Omega = \Gamma_D \cup \Gamma_N$, $\Gamma_D \cap \Gamma_N = \emptyset$, $0 < \varepsilon \ll 1$ is the diffusivity constant, $f \in L^2(\Omega)$ is the source function, $\beta \in (W^{1,\infty}(\Omega))^2$ is the velocity field, $g^D \in H^{3/2}(\Gamma_D)$ and $g^N \in H^{1/2}(\Gamma_N)$ are the Dirichlet and Neumann boundary conditions, respectively. The linear reaction coefficient α is a positive number which mimics the temporal discretization parameter $1/\Delta t$, where Δt is the time-step size. For the well-posedness of the problem, we assume also that

$$\alpha - \frac{1}{2} \nabla \cdot \beta(x) \geq \alpha_0 \geq 0 \tag{2.7}$$

for some non-negative constant α_0.

In order to discretize the problem (2.6), we apply SIPG formulation to the diffusion part given in the previous section, and the original upwinding scheme [65, 73] to the advective part. We apply the upwinding scheme by decomposing the boundary edges into the set Γ^- of inflow edges and the set Γ^+ of outflow edges defined by

$$\Gamma_h^- = \{x \in \partial \Omega : \beta \cdot \mathbf{n} < 0\}, \qquad \Gamma_h^+ = \partial \Omega \setminus \Gamma_h^-,$$

where $\mathbf{n}$ is the unit outward normal to the boundary $\partial \Omega$. The set of inflow and outflow boundary edges of an element $K \in \xi_h$ is defined in a similar way by

$$\partial K^- = \{x \in \partial K : \beta \cdot \mathbf{n}_K < 0\}, \qquad \partial K^+ = \partial K \setminus \partial K^-,$$

where $\mathbf{n}_K$ is the unit outward normal vector to the element boundary ∂K. Moreover, on an interior edge ∂K, we denote the trace of a function v from inside the element K by v^{in} and from outside the element K by v^{out}. Then, SIPG with upwinding discretized formulation of (2.6) reads as: find $u_h \in V_h$ such that

$$a_h(u_h, v_h) = l_h(v_h) \qquad \forall v_h \in V_h, \tag{2.8}$$

with the bilinear form

$$a_h(u_h, v_h) = a_h^{DR}(u_h, v_h) + a_h^C(u_h, v_h), \tag{2.9}$$

where

$$a_h^{DR}(u_h, v_h) = \sum_{K \in \xi_h} \int_K \varepsilon \nabla u_h \cdot \nabla v_h dx + \sum_{K \in \xi_h} \int_K \alpha u_h v_h dx + \sum_{e \in \Gamma_h^0 \cup \Gamma_h^D} \frac{\sigma \varepsilon}{h_e} \int_e [u_h] \cdot [v_h] ds$$

$$- \sum_{e \in \Gamma_h^0 \cup \Gamma_h^D} \int_e \{\varepsilon \nabla v_h\} \cdot [u_h] ds - \sum_{e \in \Gamma_h^0 \cup \Gamma_h^D} \int_e \{\varepsilon \nabla u_h\} \cdot [v_h] ds,$$

$$a_h^C(u_h, v_h) = \sum_{K \in \xi_h} \int_K \beta \cdot \nabla u_h v_h dx + \sum_{K \in \xi_h} \int_{\partial K^- \setminus \partial \Omega} \beta \cdot \mathbf{n}(u_h^{out} - u_h^{in}) v_h ds$$

$$- \sum_{K \in \xi_h} \int_{\partial K^- \cap \Gamma_h^-} \beta \cdot \mathbf{n} u_h^{in} v_h ds,$$

$$l_h(v_h) = \sum_{K \in \xi_h} \int_K f v_h dx + \sum_{e \in \Gamma_h^D} \int_e g^D \left(\frac{\sigma \varepsilon}{h_e} v_h - \varepsilon \nabla v_h \cdot \mathbf{n} \right) ds$$

$$- \sum_{K \in \xi_h} \int_{\partial K^- \cap \Gamma_h^-} \beta \cdot \mathbf{n} g^D v_h ds + \sum_{e \in \Gamma_h^N} \int_e g^N v_h ds.$$

Remark 2.1. Upon integration by parts on the advective term, one may also have that

$$a_h^C(u_h, v_h) = \sum_{K \in \xi_h} \int_K (-\beta u_h \cdot \nabla v_h - \nabla \cdot \beta u_h v_h) dx$$

$$+ \sum_{K \in \xi_h} \int_{\partial K^+ \setminus \partial \Omega} \beta \cdot \mathbf{n} u_h^{in} (v_h - v_h^{out}) ds + \sum_{K \in \xi_h} \int_{\partial K^+ \cap \Gamma_h^+} \beta \cdot \mathbf{n} u_h^{in} v_h ds.$$

2.5.1 Coercivity of Bilinear Form

Now, we show the coercivity of the bilinear form $a_h(u, v) = a_h^{DR}(u, v) + a_h^C(u, v)$, separately for the bilinear forms $a_h^{DR}(u, v)$ and $a_h^C(u, v)$. To show the coercivity of the bilinear form $a_h^{DR}(u, v)$, corresponding to the diffusion and linear reaction terms, we define the energy norm

$$|||v|||^2 = \sum_{K \in \xi_h} (\|\varepsilon \nabla v\|_{L^2(K)}^2 + \alpha_0 \|v\|_{L^2(K)}^2) + \sum_{e \in \Gamma_h^0 \cup \Gamma_h^D} \frac{\varepsilon \sigma}{h_e} \|[v]\|_{L^2(e)}^2. \qquad (2.10)$$

Lemma 2.1. *The bilinear form $a_h^{DR}(u, v)$ in (2.9) is coercive and satisfy*

$$a_h^{DR}(v, v) \geq \frac{1}{2} |||v|||^2, \quad \forall v \in V_h, \qquad (2.11)$$

where the norm $\|\|\| \cdot \|\|\|$ is defined as in (2.10).

Proof. The bilinear form $a_h^{DR}(u, v)$ satisfies

$$a_h^{DR}(v,v) = \sum_{K\in\xi_h} \int_K (\varepsilon(\nabla v)^2 + \alpha v^2)dx - 2\sum_{e\in\Gamma_h^0\cup\Gamma_h^D} \int_e \{\varepsilon\nabla v\}\cdot[v]ds + \sum_{e\in\Gamma_h^0\cup\Gamma_h^D} \frac{\sigma\varepsilon}{h_e}\int_e [v]^2 ds.$$

We need to find an upper bound to the term $\sum_{e\in\Gamma_h^0\cup\Gamma_h^D}\int_e\{\varepsilon\nabla v\}\cdot[v]ds$ (a lower bound to the negative of the term). Using the Cauchy-Schwarz inequality, we obtain

$$\sum_{e\in\Gamma_h^0\cup\Gamma_h^D}\int_e\{\varepsilon\nabla v\}\cdot[v]ds \le \sum_{e\in\Gamma_h^0\cup\Gamma_h^D}\|\{\varepsilon\nabla v\cdot\mathbf{n}_e\}\|_{L^2(e)}\|[v]\|_{L^2(e)}$$

$$\le \sum_{e\in\Gamma_h^0\cup\Gamma_h^D}\|\{\varepsilon\nabla v\cdot\mathbf{n}_e\}\|_{L^2(e)}\underbrace{\left(\frac{1}{|e|}\right)^{\frac{1}{2}-\frac{1}{2}}}_{1}\|[v]\|_{L^2(e)}.$$

For the interior edges $e = K_i\cap K_j \in \Gamma_h^0$, using the definition of the average operator and the trace inequality (2.1b), we get

$$\|\{\varepsilon\nabla v\cdot\mathbf{n}_e\}\|_{L^2(e)} \le \frac{1}{2}\|\varepsilon\nabla v_i\cdot\mathbf{n}_e\|_{L^2(e)} + \frac{1}{2}\|\varepsilon\nabla v_j\cdot\mathbf{n}_e\|_{L^2(e)}$$

$$\le \frac{C_{Tr}\varepsilon}{2}h_{K_i}^{-1/2}\|\nabla v\|_{L^2(K_i)} + \frac{C_{Tr}\varepsilon}{2}h_{K_j}^{-1/2}\|\nabla v\|_{L^2(K_j)}.$$

Let us denote by h the maximum element size, i.e. $h = \max(h_K)$. Obviously there holds $|e| \le h_K \le h$ for the 2D case, which of course leads to

$$\int_e\{\varepsilon\nabla v\}\cdot[v]ds \le \frac{C_{Tr}\varepsilon}{2}|e|^{1/2}\left(h_{K_i}^{-1/2}\|\nabla v\|_{L^2(K_i)} + h_{K_j}^{-1/2}\|\nabla v\|_{L^2(K_j)}\right)\left(\frac{1}{|e|}\right)^{1/2}\|[v]\|_{L^2(e)}$$

$$\le \frac{C_{Tr}\varepsilon}{2}\left(h_{K_i}^{\frac{1}{2}-\frac{1}{2}} + h_{K_j}^{\frac{1}{2}-\frac{1}{2}}\right)\left(\|\nabla v\|_{L^2(K_i)}^2 + \|\nabla v\|_{L^2(K_j)}^2\right)^{1/2}\left(\frac{1}{|e|}\right)^{1/2}\|[v]\|_{L^2(e)}$$

$$\le C_{Tr}\varepsilon\left(\|\nabla v\|_{L^2(K_i)}^2 + \|\nabla v\|_{L^2(K_j)}^2\right)^{1/2}\left(\frac{1}{|e|}\right)^{1/2}\|[v]\|_{L^2(e)}.$$

For the edges in the set of boundary edges Γ_h^D, a similar bound can be determined. Summing on $e\in\Gamma_h^0\cup\Gamma_h^D$ and noting that the maximum number of neighbors elements (triangles) in a conforming mesh is 3 results in:

$$\sum_{e\in\Gamma_h^0\cup\Gamma_h^D}\int_e\{\varepsilon\nabla v\}\cdot[v]ds \le \sqrt{3}C_{Tr}\varepsilon\left(\sum_{K\in\xi_h}\|\nabla v\|_{L^2(K)}^2\right)^{1/2}\left(\sum_{e\in\Gamma_h^0\cup\Gamma_h^D}\frac{1}{|e|}\|[v]\|_{L^2(e)}^2\right)^{1/2}.$$

For a constant $\delta > 0$, applying the Young's inequality (2.3), we obtain

$$\sum_{e\in\Gamma_h^0\cup\Gamma_h^D}\int_e\{\varepsilon\nabla v\}\cdot[v]ds \le \frac{\delta}{2}\sum_{K\in\xi_h}\|\varepsilon^{1/2}\nabla v\|_{L^2(K)}^2 + \frac{3C_{Tr}^2\varepsilon}{2\delta}\sum_{e\in\Gamma_h^0\cup\Gamma_h^D}\frac{1}{|e|}\|[v]\|_{L^2(e)}^2.$$

Now, assume that the advection field β is divergence free and the condition (2.7) gives

$$
\begin{aligned}
a_h^{DR}(v,v) &= \sum_{K \in \xi_h} \int_K (\varepsilon(\nabla v)^2 + \alpha v^2) dx - 2 \sum_{e \in \Gamma_h^0 \cup \Gamma_h^D} \int_e \{\varepsilon \nabla v\} \cdot [v] ds + \sum_{e \in \Gamma_h^0 \cup \Gamma_h^D} \frac{\sigma \varepsilon}{h_e} \int_e [v]^2 ds \\
&\geq \sum_{K \in \xi_h} \int_K (\varepsilon(\nabla v)^2 + \alpha_0 v^2) dx - 2 \sum_{e \in \Gamma_h^0 \cup \Gamma_h^D} \int_e \{\varepsilon \nabla v\} \cdot [v] ds + \sum_{e \in \Gamma_h^0 \cup \Gamma_h^D} \frac{\sigma \varepsilon}{h_e} \int_e [v]^2 ds \\
&\geq (1 - \delta) \sum_{K \in \xi_h} \|\varepsilon^{1/2} \nabla v\|_{L^2(K)}^2 + \frac{1}{2} \sum_{K \in \xi_h} \alpha_0 \|v\|_{L^2(K)}^2 + \sum_{e \in \Gamma_h^0 \cup \Gamma_h^D} \frac{\varepsilon}{|e|} \left(\sigma - \frac{3 C_{Tr}^2}{\delta} \right) \|[v]\|_{L^2(e)}^2.
\end{aligned}
$$

Finally, choosing $\delta = 1/2$ and the penalty parameter σ large enough ($\sigma \geq 6 C_{Tr}^2$) yields

$$
a_h^{DR}(v,v) \geq \frac{1}{2} |||v|||^2.
$$

To show the coercivity of the bilinear form $a_h^C(u,v)$, which corresponds to the advective part, we note that the bilinear form $a_h^C(u,v)$ is equivalent to

$$
a_h^C(u,v) = \sum_{K \in \xi_h} \int_K \beta \cdot \nabla u v \, dx + \sum_{e \in \Gamma_h^0} \int_e |\beta \cdot \mathbf{n}| (u^{in} - u^{out}) v \, ds + \sum_{e \in \Gamma_h^-} \int_e |\beta \cdot \mathbf{n}| u^{in} v \, ds.
$$

In [59], it is shown for the above definition of the bilinear form $a_h^C(u,v)$ that

$$
\begin{aligned}
a_h^C(v,v) &= \frac{1}{2} \sum_{e \in \Gamma_h^-} |\beta \cdot \mathbf{n}| \|v^{in}\|_{L^2(e)}^2 + \frac{1}{2} \sum_{e \in \Gamma_h^+} |\beta \cdot \mathbf{n}| \|v^{out}\|_{L^2(e)}^2 \\
&\quad + \frac{1}{2} \sum_{e \in \Gamma_h^0} |\beta \cdot \mathbf{n}| \|v^{in} - v^{out}\|_{L^2(e)}^2,
\end{aligned}
\tag{2.12}
$$

which yields the following lemma.

Lemma 2.2. *The bilinear form $a_h(u,v)$ (2.9) is coercive satisfying*

$$
a_h(v,v) \geq \frac{1}{2} \|v\|_{dG}^2, \quad \forall v \in V_h
$$

with the dG norm

$$
\begin{aligned}
\|v\|_{dG} &= |||v||| + \frac{1}{2} \sum_{e \in \Gamma_h^-} |\beta \cdot \mathbf{n}| \|v^{in}\|_{L^2(e)}^2 + \frac{1}{2} \sum_{e \in \Gamma_h^+} |\beta \cdot \mathbf{n}| \|v^{out}\|_{L^2(e)}^2 \\
&\quad + \frac{1}{2} \sum_{e \in \Gamma_h^0} |\beta \cdot \mathbf{n}| \|v^{in} - v^{out}\|_{L^2(e)}^2.
\end{aligned}
\tag{2.13}
$$

Proof. Using the identities (2.11) and (2.12), for all $v \in V_h$, we immediately obtain the coercivity of the bilinear form $a_h(u,v)$

$$a_h(v,v) = a_h^{DR}(v,v) + a_h^C(v,v)$$

$$\geq \frac{1}{2}|||v|||^2 + \frac{1}{2}\sum_{e\in\Gamma_h^-}|\boldsymbol{\beta}\cdot\mathbf{n}|\|v^{in}\|_{L^2(e)}^2 + \frac{1}{2}\sum_{e\in\Gamma_h^+}|\boldsymbol{\beta}\cdot\mathbf{n}|\|v^{out}\|_{L^2(e)}^2$$

$$+\frac{1}{2}\sum_{e\in\Gamma_h^0}|\boldsymbol{\beta}\cdot\mathbf{n}|\|v^{in}-v^{out}\|_{L^2(e)}^2$$

$$\geq \frac{1}{2}\|v\|_{dG}^2,$$

where the dG norm $\|\cdot\|_{dG}$ is given by (2.13).

Chapter 3
Elliptic Problems with Adaptivity

In this chapter we investigate and apply adaptive dG algorithms for the stationary semi-linear ADR equations of the model (1.1). We give the existence and uniqueness results of the elliptic system. The main focus of this chapter is handling of unphysical oscillations at the interior/boundary layers in advection dominated problems resulting through the discretization in space by applying an adaptive algorithm using residual-based robust a posteriori error estimates for the stationary model. The results obtained in this chapter for the stationary model will be a key ingredient in the follow-up chapter for the non-stationary models. Since the stiffness matrices obtained by dG methods become more dense and ill-conditioned with increasing order of dG polynomials, the resulting linear system of equations have to preconditioned. For this reason, we introduce in this chapter the matrix reordering and iterative partitioning technique in [86]. We give the details of the construction of the matrix reordering and partitioning technique, and demonstrate its efficiency numerically.

In the literature, there exist various methods to handle the sharp layers in the advection dominated ADR equations. Among them the most famous methods are the stabilized finite elements methods such as Galerkin least squares FEMs [22, 21, 58]. The most known method among this class of methods is the streamline upwind Petrov-Galerkin (SUPG) method. In this chapter, we compare the adaptive dG approximations with the Galerkin least squares FEMs in Section 3.4 and with SUPG method in Section (3.5).

3.1 Model Elliptic Problem

We consider the advection dominated scalar stationary form of the model (1.1)

© Springer International Publishing Switzerland 2016
M. Uzunca, *Adaptive Discontinuous Galerkin Methods for Non-linear Reactive Flows*,
Lecture Notes in Geosystems Mathematics and Computing,
DOI 10.1007/978-3-319-30130-3_3

$$\alpha u - \varepsilon \Delta u + \beta \cdot \nabla u + r(u) = f \qquad \text{in } \Omega, \qquad (3.1a)$$

$$u = g^D \qquad \text{on } \Gamma_D, \qquad (3.1b)$$

$$\varepsilon \frac{\partial u}{\partial \mathbf{n}} = g_N \qquad \text{on } \Gamma_N \qquad (3.1c)$$

where Ω is a bounded, open, convex domain in $\mathbb{R}^2$ with boundary $\partial \Omega = \Gamma_D \cup \Gamma_N$, $\Gamma_D \cap \Gamma_N = \emptyset$, $0 < \varepsilon \ll 1$ is the diffusivity constant, $f \in L^2(\Omega)$ is the source function, $\beta \in \left(W^{1,\infty}(\Omega) \right)^2$ is the velocity field, $g^D \in H^{3/2}(\Gamma_D)$ and $g^N \in H^{1/2}(\Gamma_N)$ are the Dirichlet and Neumann boundary conditions, respectively. The linear reaction coefficient α is a positive number which mimics the temporal discretization parameter $1/\Delta t$, where Δt is the time-step size. Further, we assume that the non-linear reaction term is bounded, locally Lipschitz continuous and monotone, i.e. satisfies for any $s, s_1, s_2 \geq 0$, $s, s_1, s_2 \in \mathbb{R}$ the following conditions

$$|r_i(s)| \leq C_S, \quad C_S > 0, \ s \in [-S, S], \qquad (3.2a)$$

$$\|r_i(s_1) - r_i(s_2)\|_{L^2(\Omega)} \leq L(S)\|s_1 - s_2\|_{L^2(\Omega)}, \quad L(S) > 0, \qquad (3.2b)$$

$$r_i \in C^1(\mathbb{R}_0^+), \quad r_i(0) = 0, \quad r_i'(s) \geq 0. \qquad (3.2c)$$

Moreover, we assume that

$$\alpha - \frac{1}{2} \nabla \cdot \beta(x) \geq \alpha_0 \geq 0, \qquad (3.3a)$$

$$\|\alpha - \nabla \cdot \beta(x)\|_{L^\infty(\Omega)} \leq c_* \alpha_0, \qquad (3.3b)$$

for some non-negative constants α_0 and c_*. The identity (3.3a) is needed to have a coercive bilinear form (well-posedness of the linear part), while, we use the identity (3.3b) to prove the reliability of the a posteriori error estimate.

In order to discretize the problem (3.1), we apply the SIPG method to the diffusion part and the upwinding scheme [65, 73] to the advective part. The solution of (3.1) reads as: find $u_h \in V_h$ such that

$$a_h(u_h, v_h) + b_h(u_h, v_h) = l_h(v_h), \qquad \forall v_h \in V_h \qquad (3.4)$$

with

$$a_h(u_h, v_h) = \sum_{K \in \xi_h} \int_K \varepsilon \nabla u_h \cdot \nabla v_h dx + \sum_{K \in \xi_h} \int_K \alpha u_h v_h dx + \sum_{K \in \xi_h} \int_K \beta \cdot \nabla u_h v_h dx$$

$$- \sum_{e \in \Gamma_h^0 \cup \Gamma_h^D} \int_e \{\varepsilon \nabla v_h\} \cdot [u_h] ds - \sum_{e \in \Gamma_h^0 \cup \Gamma_h^D} \int_e \{\varepsilon \nabla u_h\} \cdot [v_h] ds$$

$$+ \sum_{K \in \xi_h} \int_{\partial K^- \setminus \partial \Omega} \beta \cdot \mathbf{n}(u_h^{out} - u_h^{in}) v_h ds - \sum_{K \in \xi_h} \int_{\partial K^- \cap \Gamma_h^-} \beta \cdot \mathbf{n} u_h^{in} v_h ds$$

$$+ \sum_{e \in \Gamma_h^0 \cup \Gamma_h^D} \frac{\sigma \varepsilon}{h_e} \int_e [u_h] \cdot [v_h] ds,$$

$$b_h(u_h, v_h) = \sum_{K \in \xi_h} \int_K r(u_h) v_h dx,$$

$$l_h(v_h) = \sum_{K \in \xi_h} \int_K f v_h dx + \sum_{e \in \Gamma_h^D} \int_e g^D \left(\frac{\sigma \varepsilon}{h_e} v_h - \varepsilon \nabla v_h \cdot \mathbf{n} \right) ds$$

$$- \sum_{K \in \xi_h} \int_{\partial K^- \cap \Gamma_h^-} \beta \cdot \mathbf{n} g^D v_h ds + \sum_{e \in \Gamma_h^N} \int_e g^N v_h ds.$$

The formulation (3.4) differs from the one in (2.8) by the additional non-linear form $b_h(u, v)$ which is linear in the second argument. Thus, it is valid that the bilinear form $a_h(u, v)$ in (3.4) is coercive on V_h with the dG norm (2.13), which will be used to show the existence of the unique solution of the variational formulation (3.4).

3.1.1 Discrete System in Matrix-Vector Form

The SIPG discretization of the problem (3.4) has the form

$$u_h = \sum_{i=1}^{Nel} \sum_{l=1}^{Nloc} U_l^i \phi_l^i, \tag{3.5}$$

where ϕ_l^i's are the basis polynomials spanning the dGFEM space V_h, U_l^i's are the unknown coefficients which have to be computed. Nel denotes the number of triangles and $Nloc$ is the number of local dimension. We choose the piecewise basis polynomials ϕ_l^i's in such a way that each basis function has only one triangle as a support, i.e. we choose on a specific triangle K_e, $e \in \{1, 2, \ldots, Nel\}$, the basis polynomials ϕ_l^e which are zero outside the triangle K_e, $l = 1, 2, \ldots, Nloc$. By this construction, the stiffness matrix has a block structure, each block is related to a triangle or face integral (there is no overlapping as in the continuous FEM case). The product $dof := Nel \times Nloc$ gives the degree of freedom (DoFs) in dG discretization. Inserting the linear combination of u_h in (3.4) and choosing the test functions as $v_h = \phi_l^i$, $l = 1, 2, \ldots, Nloc$, $i = 1, 2, \ldots, Nel$, we obtain the non-linear system of equations in matrix-vector form

$$SU + \mathbf{b}(U) = \mathbf{L}, \tag{3.6}$$

where $\mathbf{U} \in \mathbb{R}^{dof}$ is the vector of unknown coefficients U_l^i's, $S \in \mathbb{R}^{dof \times dof}$ is the stiffness matrix corresponding to the bilinear form $a_h(u_h, v_h)$, $\mathbf{b}(U) \in \mathbb{R}^{dof}$ is the vector function of U related to the non-linear form $b_h(u_h, v_h)$ and $\mathbf{L} \in \mathbb{R}^{dof}$ is the vector to the linear form $l_h(v_h)$. They are given by

$$S = \begin{bmatrix} S_{11} & S_{12} & \cdots & S_{1,Nel} \\ S_{21} & S_{22} & & \vdots \\ \vdots & & \ddots & \\ S_{Nel,1} & \cdots & & S_{Nel,Nel} \end{bmatrix}, \quad \mathbf{U} = \begin{bmatrix} U_1 \\ U_2 \\ \vdots \\ U_{Nel} \end{bmatrix}$$

$$\mathbf{b}(U) = \begin{bmatrix} \mathbf{b}_1(U) \\ \mathbf{b}_2(U) \\ \vdots \\ \mathbf{b}_{Nel}(U) \end{bmatrix}, \quad \mathbf{L} = \begin{bmatrix} \mathbf{L}_1 \\ \mathbf{L}_2 \\ \vdots \\ \mathbf{L}_{Nel} \end{bmatrix}$$

where the block matrices are of dimension $Nloc$:

$$S_{ji} = \begin{bmatrix} a_h(\phi_1^i, \phi_1^j) & a_h(\phi_2^i, \phi_1^j) & \cdots & a_h(\phi_{Nloc}^i, \phi_1^j) \\ a_h(\phi_1^i, \phi_2^j) & a_h(\phi_2^i, \phi_2^j) & & \vdots \\ \vdots & & \ddots & \\ a_h(\phi_1^i, \phi_{Nloc}^j) & \cdots & & a_h(\phi_{Nloc}^i, \phi_{Nloc}^j) \end{bmatrix}, \quad \mathbf{U}_i = \begin{bmatrix} U_1^i \\ U_2^i \\ \vdots \\ U_{Nloc}^i \end{bmatrix}$$

$$\mathbf{b}_i = \begin{bmatrix} b_h(u_h, \phi_1^i) \\ b_h(u_h, \phi_2^i) \\ \vdots \\ b_h(u_h, \phi_{Nloc}^i) \end{bmatrix}, \quad \mathbf{L}_i = \begin{bmatrix} l_h(\phi_1^i) \\ l_h(\phi_2^i) \\ \vdots \\ l_h(\phi_{Nloc}^i) \end{bmatrix}$$

Proposition 3.1. *The non-linear vector* $\mathbf{b}(U)$ *in* (3.6) *is locally Lipschitz with respect to* $\mathbf{U}$.

Proof. For given functions u^1, u^2, v with u^1, u^2, satisfying (3.5), we have by definition

$$b_h(u^1 - u^2, v) = \int_\Omega r(u^1 - u^2)v dx.$$

Applying Cauchy-Schwarz's inequality and using the locally Lipschitz continuity condition (3.2b), we get

$$b_h(u^1 - u^2, v) \leq \|r(u^1 - u^2)\|_{L^2(\Omega)} \|v\|_{L^2(\Omega)}$$
$$\leq L_S \|u^1 - u^2\|_{L^2(\Omega)} \|v\|_{L^2(\Omega)},$$

which means that the non-linear form $b_h(u, v)$ is locally Lipschitz continuous in the first argument. Since the components of the vector $\mathbf{b}(U^1 - U^2)$ are nothing but the non-linear forms $b_h(u^1 - u^2, \phi_l^i)$, $l = 1, 2, \ldots, Nloc$, $i = 1, 2, \ldots, Nel$, each component of the vector $\mathbf{b}(\mathbf{U})$ is locally Lipschitz continuous, which yields that the vector $\mathbf{b}(\mathbf{U})$ is locally Lipschitz with respect to $\mathbf{U}$.

Proposition 3.2 (Existence and uniqueness of the discrete solutions). *The SIPG formulation* (3.4) *has a unique solution.*

Proof. The coercivity (Lemma 2.2) of the bilinear form $a_h(u, v)$ implies that the stiffness matrix S of (3.6) arising from the bilinear form $a_h(u, v)$ is positive definite. Combining this with the locally Lipschitz condition (Lemma 3.1) of the non-linear vector $\mathbf{b}(\mathbf{U})$ in (3.6), we prove that the SIPG discretization (3.4) produces a unique solution.

3.2 Adaptivity

Advection dominated problems lead to internal/boundary layers where the solution has large gradients. The standard FEMs are known to produce strong oscillations around the layers. A naive approach is to refine the mesh uniformly. But it is not desirable as it highly increase the degree of freedom and refines the mesh unnecessarily in regions where the solutions are smooth. On the other hand, in the semi-linear ADR equations, in addition to the nonphysical oscillations due to the advection, non-linear reaction can be also responsible for sharp fronts. Only adaptive discretization methods can overcome all these nonphysical oscillations and shocks, where the mesh is refined only locally.

In adaptive algorithms the elements in a triangulation are selected and refined locally when the estimated local errors are large. Thus, the crucial part of an adaptive algorithm is to estimate the local errors. The major tool to estimate the local errors is the a posteriori error estimation using the approximate solution and the given problem data. There are many studies on a posteriori error estimation most of them based on the energy norm induced by the weak formulation [3, 10, 93, 92, 91]. On the other hand, the local structure of the dG methods make them suitable for adaptive schemes. The convergence analysis of a residual-based a posteriori error estimation using dG was first studied by Karakashian and Pascal [61]. Hoppe et al. [53] analyzed the convergence of the a posteriori error estimates for the IPdG method. In [2], a posteriori error estimator robust in any unknown constant has been derived by Ainsworth. Further, a posteriori error estimation using dG discretization are also studied by Rivière et al. [76], Houston et al. [56] and Ern et al. [43], and references therein.

In this section, we introduce an adaptive strategy for elliptic semi-linear ADR equations using the residual-based robust a posteriori error estimators for advection diffusion equations in [80]. We prove the a posteriori bounds with respect to the energy norm induced by the SIPG discretization.

3.2.1 The Adaptive Algorithm

Our adaptive algorithm is based on the standard adaptive finite element (AFEM) iterative loop (Fig.3.1):

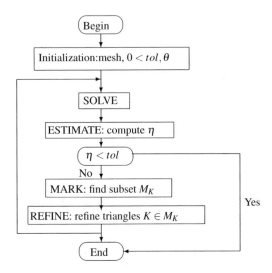

Fig. 3.1: Adaptive strategy

The first step, SOLVE, is to solve the SIPG discretized system (3.4) on a given triangulation ξ_h. The ESTIMATE step is the key part of the adaptive procedure, by which the elements with large error are selected to be refined. As an estimator, we use the modified version of the residual-based error indicator and of the error estimator in Schötzau and Zhu [80]. We add to the a posteriori error indicator a non-linear reaction term as a local contribution to the cell residuals and not to the interior/boundary edge residuals [94, Chp. 5.1.4]. Let u_h be the solution to (3.4). Then, for each element $K \in \xi_h$, we define the local error indicators η_K^2 as

$$\eta_K^2 = \eta_{R_K}^2 + \eta_{E_K^0}^2 + \eta_{E_K^D}^2 + \eta_{E_K^N}^2. \tag{3.7}$$

In (3.7), η_{R_K} denote the cell residuals

$$\eta_{R_K}^2 = \rho_K^2 \|f - \alpha u_h + \varepsilon \Delta u_h - \beta \cdot \nabla u_h - r(u_h)\|_{L^2(K)}^2,$$

while, $\eta_{E_K^0}$, $\eta_{E_K^D}$ and $\eta_{E_K^N}$ stand for the edge residuals coming from the jump of the numerical solution on the interior, Dirichlet boundary and Neumann boundary edges, respectively,

$$\eta_{E_K^0}^2 = \sum_{e \in \partial K \cap \Gamma_h^0} \left(\frac{1}{2} \varepsilon^{-\frac{1}{2}} \rho_e \| [\varepsilon \nabla u_h] \|_{L^2(e)}^2 + \frac{1}{2} \left(\frac{\varepsilon \sigma}{h_e} + \alpha_0 h_e + \frac{h_e}{\varepsilon} \right) \| [u_h] \|_{L^2(e)}^2 \right),$$

$$\eta_{E_K^D}^2 = \sum_{e \in \partial K \cap \Gamma_h^D} \left(\frac{\varepsilon \sigma}{h_e} + \alpha_0 h_e + \frac{h_e}{\varepsilon} \right) \| g^D - u_h \|_{L^2(e)}^2,$$

$$\eta_{E_K^N}^2 = \sum_{e \in \partial K \cap \Gamma_h^N} \varepsilon^{-\frac{1}{2}} \rho_e \| g^N - \varepsilon \nabla u_h \cdot \mathbf{n} \|_{L^2(e)}^2,$$

where the weights ρ_K and ρ_e, on an element K and along an edge e, respectively, are defined by

$$\rho_K = \min\{h_K \varepsilon^{-\frac{1}{2}}, \alpha_0^{-\frac{1}{2}}\}, \quad \rho_e = \min\{h_e \varepsilon^{-\frac{1}{2}}, \alpha_0^{-\frac{1}{2}}\},$$

for $\alpha_0 \neq 0$. When $\alpha_0 = 0$, we take $\rho_K = h_K \varepsilon^{-\frac{1}{2}}$ and $\rho_e = h_e \varepsilon^{-\frac{1}{2}}$. Then, our a posteriori error indicator is given by

$$\eta = \left(\sum_{K \in \xi_h} \eta_K^2 \right)^{1/2}. \tag{3.8}$$

We also introduce the data approximation error,

$$\Theta^2 = \Theta^2(f) + \Theta^2(u^D) + \Theta^2(u^N) \tag{3.9}$$

with

$$\Theta^2(f) = \sum_{K \in \xi} \rho_K^2 (\| f - f_h \|_{L^2(K)}^2 + \| (\beta - \beta_h) \cdot \nabla u_h \|_{L^2(K)}^2 + \| (\alpha - \alpha_h) u_h \|_{L^2(K)}^2),$$

$$\Theta^2(u^D) = \sum_{e \in \Gamma_h^D} \left(\frac{\varepsilon \sigma}{h_e} + \alpha_0 h_e + \frac{h_e}{\varepsilon} \right) \| g^D - \hat{g}^D \|_{L^2(e)}^2,$$

$$\Theta^2(u^N) = \sum_{e \in \Gamma_h^N} \varepsilon^{-\frac{1}{2}} \rho_e \| g^N - \hat{g}^N \|_{L^2(e)}^2$$

with $\hat{g}^D$ and $\hat{g}^N$ denoting the mean integrals of g^D and g^N, respectively.

In the MARK step, if the given tolerance is not satisfied, we determine the set of elements $M_K \subset \xi_h$ to be refined using the error indicator (3.7). This is handled using the bulk criterion proposed by Döfler [39], by which the approximation error is decreased by a fixed factor for each loop. We choose the set of elements $M_K \subset \xi_h$ satisfying

$$\sum_{K \in M_K} \eta_K^2 \geq \theta \sum_{K \in \xi_h} \eta_K^2$$

for a user defined parameter $0 < \theta < 1$. Here, bigger θ results in increasing number of refinement of triangles in a single loop, where, smaller θ causes more refinement loops.

Finally, in the REFINE step, we refine the marked elements $K \in M_K$ using the newest vertex bisection method [29]. This process can be summarized as (see Fig.3.2): for each element $K \in \xi_h$, we label one vertex of K as a newest vertex. The opposite edge of the newest vertex is called the refinement edge. Then, a triangle is bisected to two new children triangles by connecting the newest vertex to the midpoint of the refinement edge, and this new vertex created at the midpoint of the refinement edge is assigned to be the newest vertex of the children. Following a similar rule, these two children triangles are bisected to obtain four children elements belonging to the father element (the refined triangle $K \in M_K$). After bisecting all $K \in M_K$, we also divide some elements $K \in \xi_h \setminus M_K$ to keep the conformity of the mesh, i.e. hanging nodes are not allowed.

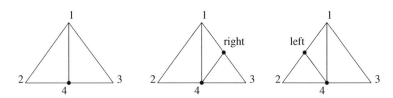

Fig. 3.2: Bisection of a triangle

In the case of coupled problems, instead of a single component problem, we refine the elements in the union of the set of the elements to be refined for each component, i.e., let η_K^1 and η_K^2 be the computed local error indicators corresponding to each unknown component of a two component system. Next, we determine the set of elements M_K^1 and M_K^2 satisfying

$$\sum_{K \in M_K^1} (\eta_K^1)^2 \geq \theta \sum_{K \in \xi_h} (\eta_K^1)^2 \,, \qquad \sum_{K \in M_K^2} (\eta_K^2)^2 \geq \theta \sum_{K \in \xi_h} (\eta_K^2)^2.$$

Then, we refine the marked elements $K \in M_K^1 \cup M_K^2$ using the newest vertex bisection method. The adaptive procedure ends after a sequence of mesh refinements to attain a solution within a prescribed tolerance.

3.2.2 A Posteriori Error Estimation

We use the following energy norm to measure the a posteriori error:

$$|||v|||^2 = \sum_{K \in \xi_h} (\|\varepsilon \nabla v\|^2_{L^2(K)} + \alpha_0 \|v\|^2_{L^2(K)}) + \sum_{e \in \Gamma_h^0 \cup \Gamma_h^D} \frac{\varepsilon \sigma}{h_e} \|[v]\|^2_{L^2(e)}, \qquad (3.10)$$

and the semi-norm

$$|v|^2_C = |\beta v|^2_* + \sum_{e \in \Gamma_h^0 \cup \Gamma_h^D} (\alpha_0 h_e + \frac{h_e}{\varepsilon}) \|[v]\|^2_{L^2(e)}, \qquad (3.11)$$

where

$$|u|_* = \sup_{w \in H^1_0(\Omega) \setminus \{0\}} \frac{\int_\Omega u \cdot \nabla w \, dx}{|||w|||}.$$

The terms $|\beta v|^2_*$ and $\frac{h_e}{\varepsilon} \|[v]\|^2_{L^2(e)}$ in (3.11) are used to bound the advective part, whereas, the term $\alpha_0 h_e \|[v]\|^2_{L^2(e)}$ is used to bound the linear reaction part of the discrete system.

Theorem 3.1. *Let u and u_h be the solutions of the continuous problem (3.1) and the discrete SIPG problem (3.4), respectively. We assume that (3.2) and (3.3) hold. Then, we obtain the a posteriori error estimates*

$$|||u - u_h||| + |u - u_h|_C \lesssim \eta + \Theta \qquad \text{(reliability)}, \qquad (3.12)$$

$$\eta \lesssim |||u - u_h||| + |u - u_h|_C + \Theta \qquad \text{(efficiency)}, \qquad (3.13)$$

where η is the error indicator and Θ the data oscillation error, defined in (3.7) and (3.9), respectively.

Before continuing the proof of the Theorem 3.1, let us first introduce some key tools used in the proof. The proof of the Theorem 3.1 is analogous to the proof for linear problems in [80]. Therefore, we give the proof for the non-linear reaction term only. In the following, we use the symbols $\lesssim$ and $\gtrsim$ to denote the bounds that are valid up to positive constants independent of the local mesh size h, the diffusion coefficient ε and the penalty parameter σ. Further, we use the dG norm defined by

$$\|v\|^2_{dG} = |||v||| + |v|_C \qquad (3.14)$$

with the definitions in (3.10) and (3.11).

We note that u and u_h denote the solutions of the continuous (3.1) and the discrete problems (3.4), respectively. Because the error $\|u - u_h\|_{dG}$ is not well-defined, since $u \in H^1_0(\Omega)$ and $u_h \in V_h \not\subseteq H^1_0(\Omega)$. Therefore, we split SIPG solution u_h as

$$u_h = u_h^c + u_h^r$$

with $u_h^c \in H^1_0(\Omega) \cap V_h$ denoting the conforming part of the solution and $u_h^r \in V_h$ is the remainder term. Therefore $u_h \in H^1_0(\Omega) + V_h$, and from the triangular inequality, we obtain

$$\|u - u_h\|_{dG} \leq \|u - u_h^c\|_{dG} + \|u_h^r\|_{dG}.$$

Now, both terms on the right-hand side are well-defined, which have to be evaluated. Next, using the Remark 2.1 for the advective term, we introduce the following auxiliary forms:

$$D_h(u,v) = \sum_{K \in \xi_h} \int_K (\varepsilon \nabla u \cdot \nabla v + (\alpha - \nabla \cdot \beta) uv) \, dx, \tag{3.15a}$$

$$O_h(u,v) = - \sum_{K \in \xi_h} \int_K \beta u \cdot \nabla v \, dx + \sum_{K \in \xi_h} \int_{\partial K^+ \cap \Gamma^+} \beta \cdot \mathbf{n}_K uv \, ds$$

$$+ \sum_{K \in \xi_h} \int_{\partial K^+ \setminus \partial \Omega} \beta \cdot \mathbf{n}_K u(v - v^{out}) \, ds, \tag{3.15b}$$

$$K_h(u,v) = - \sum_{e \in \Gamma_h^0 \cup \Gamma_h^D} \int_e \{\varepsilon \nabla u\} \cdot [v] \, ds - \sum_{e \in \Gamma_h^0 \cup \Gamma_h^D} \int_e \{\varepsilon \nabla v\} \cdot [u] \, ds, \tag{3.15c}$$

$$J_h(u,v) = \sum_{e \in \Gamma_h^0 \cup \Gamma_h^D} \frac{\sigma \varepsilon}{h_e} \int_e [u] \cdot [v] \, ds. \tag{3.15d}$$

We define the bilinear form $\tilde{a}_h(u,v)$ by

$$\tilde{a}_h(u,v) = D_h(u,v) + O_h(u,v) + J_h(u,v),$$

which is well-defined on $H_0^1(\Omega) + V_h$ and is coercive [80, Lemma 4.1]

$$\tilde{a}_h(v,v) \geq |||v|||^2, \quad \forall v \in H_0^1(\Omega).$$

Moreover, the SIPG bilinear form $a_h(u,v)$ in (3.4) satisfies

$$a_h(u,v) = \tilde{a}_h(u,v) + K_h(u,v) \qquad \forall u,v \in V_h, \tag{3.16}$$

$$a_h(u,v) = \tilde{a}_h(u,v) \qquad \forall u,v \in H_0^1(\Omega). \tag{3.17}$$

Further, the auxiliary forms are continuous [80, Lemma 4.2]:

$$|D_h(u,v)| \lesssim |||u||| \, |||v||| \qquad u,v \in H_0^1(\Omega) + V_h, \tag{3.18}$$

$$|O_h(u,v)| \lesssim |\beta u|_* \, |||v||| \qquad u \in H_0^1(\Omega) + V_h, v \in H_0^1(\Omega), \tag{3.19}$$

$$|J_h(u,v)| \lesssim |||u||| \, |||v||| \qquad u,v \in H_0^1(\Omega) + V_h, \tag{3.20}$$

and for $u \in V_h$, $v \in V_h \cap H_0^1(\Omega)$, we have [80, Lemma 4.3]

$$|K_h(u,v)| \lesssim \sigma^{-1/2} \left(\sum_{e \in \Gamma_h^0 \cup \Gamma_h^D} \frac{\sigma \varepsilon}{h_e} \| [u] \|_{L^2(e)} \right)^{1/2} |||v|||. \tag{3.21}$$

We also have for the non-linear form $b_h(u,v)$, using the assumption (3.2a)

$$|b_h(u,v)| \lesssim |||v|||, \qquad u,v \in H_0^1(\Omega)+V_h. \tag{3.22}$$

Now, we give some auxiliary results and conditions to be used in the proof of the Theorem 3.1.

Lemma 3.1. *[80, Lemma 4.4](Inf-sup condition) For all $u \in H_0^1(\Omega)$, we have*

$$|||u||| + |\beta u|_* \lesssim \sup_{v \in H_0^1(\Omega)\setminus\{0\}} \frac{\tilde{a}_h(t;u,v)}{|||v|||}. \tag{3.23}$$

Definition 3.1. (Approximation operator) Let $V_h^c = V_h \cap H_0^1(\Omega)$ be the conforming subspace of V_h. For any $u \in V_h$, there exists an approximation operator $A_h : V_h \mapsto V_h^c$ satisfying

$$\sum_{K \in \xi} \|u - A_h u\|_{L^2(K)}^2 \lesssim \sum_{e \in \Gamma_h^0 \cup \Gamma_h^D} \int_e h_e |[u]|^2 ds, \tag{3.24}$$

$$\sum_{K \in \xi} \|\nabla(u - A_h u)\|_{L^2(K)}^2 \lesssim \sum_{e \in \Gamma_h^0 \cup \Gamma_h^D} \int_e \frac{1}{h_e} |[u]|^2 ds. \tag{3.25}$$

Definition 3.2. (Interpolation operator) For any $u \in H_0^1(\Omega)$, there exists an interpolation operator

$$I_h : H_0^1(\Omega) \mapsto \{w \in C(\overline{\Omega}) : w|_K \in \mathbb{P}_1(K), \forall K \in \xi, w = 0 \text{ on } \Gamma\},$$

which satisfies

$$|||I_h u||| \lesssim |||u|||, \tag{3.26}$$

$$\left(\sum_{K \in \xi} \rho_K^{-2} \|u - I_h u\|_{L^2(K)}^2\right)^{1/2} \lesssim |||u|||, \tag{3.27}$$

$$\left(\sum_{e \in \Gamma_0 \cup \Gamma^D} \varepsilon^{1/2} \rho_e^{-1} \|u - I_h u\|_{L^2(K)}^2\right)^{1/2} \lesssim |||u|||. \tag{3.28}$$

Now, consider the splitting of the discrete solution $u_h = u_h^c + u_h^r$ as $u_h^c = A_h u_h \in H_0^1(\Omega) \cap V_h$ with A_h is the approximation operator in Definition 3.1 and $u_h^r = u_h - u_h^c \in V_h$.

Lemma 3.2. *[80, Lemma 4.7](Bound for the remainder term) For the remainder term, we have the bound*

$$\|u_h^r\|_{dG} \lesssim \eta, \tag{3.29}$$

where η is the a posteriori error indicator (3.8).

Lemma 3.3. *For any $v \in H_0^1(\Omega)$, we have*

$$\int_\Omega f(v - I_h v)dx - \tilde{a}_h(u_h, v - I_h v) - b_h(u_h, v - I_h v) \lesssim (\eta + \Theta)|||v|||, \qquad (3.30)$$

where I_h is the interpolation operator in Definition 3.2.

Proof. Let

$$T = \int_\Omega f(v - I_h v)dx - \tilde{a}_h(u_h, v - I_h v) - b_h(u_h, v - I_h v).$$

Applying integration by parts, we get

$$T = \sum_{K \in \xi_h} \int_K (f - \alpha u_h + \varepsilon \Delta u_h - \beta \cdot \nabla u_h - r(u_h))(v - I_h v)dx$$

$$- \sum_{K \in \xi_h} \int_{\partial K} \varepsilon \nabla u_h \cdot \mathbf{n}_K (v - I_h v)ds$$

$$+ \sum_{K \in \xi_h} \int_{\partial K^- \setminus \partial \Omega} \beta \cdot \mathbf{n}_K (u_h - u_h^{out})(v - I_h v)ds$$

$$= T_1 + T_2 + T_3.$$

Adding and subtracting the data approximation terms into the term T_1, we obtain

$$T_1 = \sum_{K \in \xi_h} \int_K (f_h - \alpha u_h + \varepsilon \Delta u_h - \beta_h \cdot \nabla u_h - r(u_h))(v - I_h v)dx$$

$$+ \sum_{K \in \xi_h} \int_K ((f - f_h) - (\alpha - \alpha_h)u_h - (\beta - \beta_h) \cdot \nabla u_h)(v - I_h v)dx.$$

Using the Cauchy-Schwarz inequality and interpolation operator identity (3.27) gives

$$T_1 \lesssim \left(\sum_{K \in \xi_h} \eta_{R_K}^2 \right)^{1/2} \left(\sum_{K \in \xi_h} \rho_K^{-2} \|v - I_h v\|_{L^2(K)}^2 \right)^{1/2}$$

$$+ \left(\sum_{K \in \xi_h} \Theta_K^2 \right)^{1/2} \left(\sum_{K \in \xi_h} \rho_K^{-2} \|v - I_h v\|_{L^2(K)}^2 \right)^{1/2}$$

$$\lesssim \left(\sum_{K \in \xi_h} (\eta_{R_K}^2 + \Theta_K^2) \right)^{1/2} |||v|||.$$

Moreover, for the terms T_2 and T_3, we have [80, Lemma 4.8]

$$T_2 \lesssim \left(\sum_{K \in \xi_h} \eta_{E_K}^2 \right)^{1/2} |||v|||,$$

$$T_3 \lesssim \left(\sum_{K \in \xi_h} \eta_{J_K}^2 \right)^{1/2} |||v|||,$$

which completes the proof.

Lemma 3.4. *(Bound to the conforming part of the error) The conforming part of the error satisfies*

$$\|u - u_h^c\|_{dG} \lesssim \eta + \Theta. \tag{3.31}$$

Proof. Since $u - u_h^c \in H_0^1(\Omega)$, we have $|u - u_h^c|_C = |\beta(u - u_h^c)|_*$. Then, from the inf-sup condition (3.23), we get

$$\|u - u_h^c\|_{dG} = |||u - u_h^c||| + |u - u_h^c|_C \lesssim \sup_{v \in H_0^1(\Omega) \backslash \{0\}} \frac{\tilde{a}_h(u - u_h^c, v)}{|||v|||}.$$

So, we need to bound the term $\tilde{a}_h(u - u_h^c, v)$. Using the fact that $u - u_h^c \in H_0^1(\Omega)$, we have

$$\tilde{a}_h(u - u_h^c, v) = \tilde{a}_h(u, v) - \tilde{a}_h(u_h^c, v)$$

$$= \int_\Omega f v dx - b_h(u, v) - \tilde{a}_h(u_h^c, v)$$

$$= \int_\Omega f v dx - b_h(u, v) - D_h(u_h^c, v) - J_h(u_h^c, v) - O_h(u_h^c, v)$$

$$= \int_\Omega f v dx - b_h(u_h, v) + b_h(u_h, v) - b_h(u, v)$$
$$- \tilde{a}_h(u_h, v) + D_h(u_h^r, v) + J_h(u_h^r, v) + O_h(u_h^r, v).$$

We also have from the SIPG scheme (3.4) that

$$\int_\Omega f I_h v dx = \tilde{a}_h(u_h, I_h v) + K_h(u_h, I_h v) + b_h(u_h, I_h v),$$

where I_h is the interpolation operator in Definition 3.2. Hence, we obtain

$$\tilde{a}(u - u_h^c, v) = T_1 + T_2 + T_3 + T_4,$$

for which

$$T_1 = \int_\Omega f(v - I_h v) dx - \tilde{a}_h(u_h, v - I_h v) - b_h(u_h, v - I_h v)$$
$$T_2 = D_h(u_h^r, v) + J_h(u_h^r, v) + O_h(u_h^r, v)$$
$$T_3 = K_h(u_h^s, I_h v)$$
$$T_4 = b_h(u_h, v) - b_h(u, v).$$

From the identity (3.30), we have

$$T_1 \lesssim (\eta + \Theta)|||v|||.$$

The continuity results (3.18)-(3.20) and the bound for the remainder (3.29) yields

$$T_2 \lesssim (|||u_h^r||| + |\beta u_h^r|_*)|||v||| \leq \eta |||v|||.$$

Moreover, using the identities (3.21) and (3.26), we get

$$T_3 \lesssim \sigma^{-1/2} \left(\sum_{K \in \xi} \eta_{J_K}^2 \right)^{1/2} |||I_h v||| \lesssim \sigma^{-1/2} \left(\sum_{K \in \xi} \eta_{J_K}^2 \right)^{1/2} |||v|||.$$

Finally, using the Cauchy-Schwarz inequality and the identity 3.22, we obtain

$$T_4 = b_h(u_h, v) - b_h(u, v) = \int_\Omega r(u_h)v dx - \int_\Omega r(u)v dx$$
$$\leq C_1 \|v\|_{L^2(\Omega)} - C_2 \|v\|_{L^2(\Omega)}$$
$$\lesssim |||v|||,$$

which completes the proof.

Now, we can give the proof of Theorem 3.1.

Proof. (Theorem 3.1) In case of the reliability condition (3.12), combining the bounds (3.29) and (3.31) for the remainder and the conforming parts of the error, respectively, we obtain

$$\|u - u_h\|_{dG} \leq \|u - u_h^c\|_{dG} + \|u_h^r\|_{dG}$$
$$\leq \eta + \Theta + \eta$$
$$\lesssim \eta + \Theta.$$

The proof of the efficiency condition (3.13) is similar to [80, Theorem 3.3]. We only use the boundedness property (3.2a) of the non-linear reaction term to bound the terms occurring in the procedure in [80].

3.3 Solution of Linearized Systems

The approximate dG solution u_h in Section 3.1.1 has the form

$$u_h = \sum_{i=1}^{Nel} \sum_{l=1}^{Nloc} U_l^i \phi_l^i,$$

where ϕ_l^i's are the basis polynomials spanning the dGFEM space V_h, U_l^i's are the unknown coefficients, Nel denotes the number of triangles and $Nloc$ is the number of local dimension depending on the degree of polynomials k. The dimension of the system is $dof := Nel \times Nloc$ known as the degree of freedoms (DoFs) of dG approximation. Inserting the linear combination of u_h in (3.4) and choosing the test functions as $v_h = \phi_l^i$, $l = 1, 2, \ldots, Nloc$, $i = 1, 2, \ldots, Nel$, the discrete residual of the system (3.4) in matrix vector form is given by

$$R(\mathbf{U}) = S\mathbf{U} + \mathbf{b}(\mathbf{U}) - \mathbf{L} = 0, \tag{3.32}$$

where $\mathbf{U} \in \mathbb{R}^{dof}$ is the vector of unknown coefficients U_l^i's, and the stiffness matrix $S \in \mathbb{R}^{dof \times dof}$, the vector function $h \in \mathbb{R}^{dof}$ and the vector $\mathbf{L} \in \mathbb{R}^{dof}$ are defined as in Section 3.1.1. We apply the Newton-Raphson method to solve the non-linear system of equations (3.32). The Newton-Raphson solution of the system (3.32) reads as: given initial guess $\mathbf{U}^{(0)}$, for $i = 0, 1, 2, \ldots$, solve the system

$$\begin{aligned} Jw^{(i)} &= -R^{(i)}, \\ \mathbf{U}^{(i+1)} &= \mathbf{U}^{(i)} + w^{(i)} \end{aligned} \tag{3.33}$$

until a prescribed tolerance is satisfied. In (3.33), the sparse matrix $J = S + \mathbf{b}'(\mathbf{U}^{(0)})$ denotes the value of the Jacobian matrix of the residual $R(\mathbf{U})$ at the initial iterate $\mathbf{U}^{(0)}$ and remains unchanged among the iteration steps, $w^{(i)} = \mathbf{U}^{(i+1)} - \mathbf{U}^{(i)}$ is the Newton correction, and $R^{(i)} = R(\mathbf{U}^{(i)})$ denotes the value of the residual $R(\mathbf{U})$ at the current iterate $\mathbf{U}^i$.

The solution of the large linear system (3.33) is challenging since stiffness matrices are ill-conditioned and dense for higher order dG elements [7]. Therefore the linear system of (3.33) should be solved using a preconditioner. There are several preconditioners developed for the efficient and accurate solution of the linear advection-diffusion equations [4, 48]. In this section, we introduce the matrix reordering and partitioning technique in [86] as a preconditioner.

3.3.1 Matrix Reordering & Block LU Factorization

We apply the matrix reordering and partitioning technique in [86], which uses the largest eigenvalue and corresponding eigenvector of the Laplacian matrix. This reordering allows partitioning and a preconditioner based on this partitioning. Since our matrices arising from the SIPG formulation are non-symmetric due to the advective terms, in the first step, we compute the symmetric matrices by adding its transpose to itself and taking the average. A symmetric, square and sparse matrix could be represented as a graph where same index rows and columns are mapped into vertices and nonzero entries are mapped into the edges of the graph. Because the matrix is symmetric, the resulting graph is undirected. The Laplacian matrix ($\mathscr{L}$) is, then, defined as follows

$$\mathscr{L}(i, j) = \begin{cases} deg(v_i) & \text{if } i = j, \\ -1 & \text{if } i \neq j \end{cases}$$

in which the $deg(v_i)$ is the degree of the vertex i. We use the reordering based on the unweighted Laplacian matrix given above. If the graph contains only one connected component, the eigenvalues of $\mathcal{L}$ are $0 = \lambda_1 < \lambda_2 \leq \lambda_3 \leq ... \leq \lambda_n$, otherwise there are as many zero eigenvalues as the number of connected components.

Certain eigenvalues and corresponding eigenvectors of the Laplacian matrix have been studied extensively. Most notably the second nontrivial eigenvalue of the Laplacian and the corresponding eigenvector known as the algebraic connectivity and the Fiedler vector of the graph [45]. The Nodal domain theorem in [46] shows that the eigenvectors corresponding to the eigenvalues other than the first and the second smallest eigenvalue give us the connected components of the graph. In [11], the Fiedler vector for permuting the matrices to reduce the bandwidth is proposed. Reordering to obtain effective and scalable parallel banded preconditioners is proposed in [68]. We use a sparse matrix reordering for partitioning and solving linear systems using the largest eigenvalue and the corresponding eigenvector of the Laplacian matrix. Using this reordering, we show that one can reveal underlying structure of a sparse matrix. A simple Matlab code to find the reordered matrix and the permutation matrix can be found in Appendix A.

The preconditioned linear system (3.33) is formed by constructing a permutation matrix P applying the matrix reordering technique described above to the sparse Jacobian matrix J. Then, we apply the permutation matrix P to obtain the permuted system $Nw = d$ where $N = PJP^T$, $w = Pw^{(i)}$ and $d = -PR^{(i)}$. After solving the permuted system, the solution of the unpermuted linear system (3.33) can be obtained by applying the inverse permutation, $w^{(i)} = P^T w$.

Given a sparse linear system of equations $Nw = d$, after reordering, one way is to solve this system is via block LU factorization. Denoting the permuted matrix as N, the right-hand side d and the solution w is partitioned as follows:

$$\begin{pmatrix} A & B \\ C^T & D \end{pmatrix} \begin{pmatrix} w_1 \\ w_2 \end{pmatrix} = \begin{pmatrix} d_1 \\ d_2 \end{pmatrix}.$$

A block LU factorization of the coefficient matrix can be given as

$$\begin{pmatrix} A & B \\ C^T & D \end{pmatrix} = \begin{pmatrix} A & 0 \\ C^T & S \end{pmatrix} \begin{pmatrix} I & U \\ 0 & I \end{pmatrix},$$

where $U = A^{-1}B$ and $S = D - C^T A^{-1} B$, is the Schur complement matrix. The computational cost is reduced by forming the matrices U and S once and using them for solving linear systems with the same coefficient matrix and different right-hand sides. After this factorization, there are different possibilities to solve the system. One approach is to solve the system via block backward and forward substitution, by first solving the linear system $At = d_1$, and then solving the Schur complement system $Sw_2 = d_2 - C^T t$ and obtaining $w_1 = t - Uw_2$. This method is summarized in Algorithm 1.

We note that this approach involves solving two linear systems of equations with the coefficient matrix A and S. These linear systems can be solved directly or iteratively with effective preconditioners. Other approaches could involve solving the

---

Algorithm 1 Algorithm for solving the linear system after reordering

---

Input: The coefficient matrix: $\begin{pmatrix} A & B \\ C^T & D \end{pmatrix}$ and the right-hand side: $\begin{pmatrix} d_1 \\ d_2 \end{pmatrix}$

Output: The solution vector: $\begin{pmatrix} w_1 \\ w_2 \end{pmatrix}$

1: solve $At = d_1$
2: solve $Sw_2 = d_2 - C^T t$
3: compute $w_1 = t - U w_2$

---

system $Nw = d$ iteratively where the preconditioner may take other forms. There are many other techniques for solving block partitioned and saddle point linear systems, we refer to [19] for a more detailed survey of some of these methods.

3.4 Comparison with Galerkin Least Squares FEM (GLSFEM)

One of the most popular methods for solving advection dominated ADR equation is the Galerkin Least Squares FEM (GLSFEM) such as SUPG. In this section we discuss GLSFEMs and compare with dGFEM and adaptive dGFEM (dGAFEM).

For linear PDEs, the weak form in the standard Galerkin method is obtained by multiplying the differential equation with a test function v and integrating over a suitable function space V

$$(\mathscr{L}u, v) = (f, v), \quad \forall v \in V = H_0^1(\Omega),$$

where $\mathscr{L} = -\varepsilon \Delta + \beta \cdot \nabla + \alpha$ is the linear part of the stationary advection-diffusion-reaction equation (3.1). Defining the residual as $R(u) = f - \mathscr{L}u$, the standard Galerkin method can be interpreted in the form of the orthogonality of the residual $(R(u), v) = 0$. In the case of non self-adjoint differential operators $\mathscr{L}$ like in the ADR equations, it can happen that $(\mathscr{L}u, v)$ is not coercive or not symmetric on V, and the resulting FEM discretization becomes unstable.

For transport problems, another popular approach is based on the least squares formulation of the Galerkin FEM. Let us write the simple form of the model problem (3.1) as

$$\mathscr{L}u + r(u) = f \quad \text{in } \Omega, \tag{3.34a}$$
$$u = g \quad \text{on } \partial\Omega. \tag{3.34b}$$

Define the least-squares functional

$$J(u) := \frac{1}{2} \|\mathscr{L}u + r(u) - f\|_{L^2(\Omega)}^2.$$

A minimizer of $J(u)$ is obtained by

$$\lim_{t \to 0} \frac{d}{dt} J(u+tv) = 0 , \quad \forall v \in V,$$

which yields the least-squares term

$$\tilde{J}_\Omega(u,v) := (\mathscr{L}u + r(u) - f, \mathscr{L}v + r'(u)v)_{L^2(\Omega)}.$$

For linear problems with $r(u) = 0$, the least squares Galerkin method reduces to the minimization problem

$$F(u) = \min_{v \in V} F(v),$$

where the functional $F(\cdot)$ is defined by

$$F(v) = \frac{1}{2} \| \mathscr{L}v - f \|^2_{L^2(\Omega)} .$$

The first-order optimality condition leads to the least squares Galerkin method

$$(\mathscr{L}u, \mathscr{L}v) = (f, \mathscr{L}v) , \quad \forall v \in V.$$

The bilinear form $(\mathscr{L}u, \mathscr{L}v)$ is symmetric and coercive and has stronger stability properties compared to the standard Galerkin method.

There are many publications on the Galerkin least squares finite element methods (GLSFEM). We mention here two books [22, 58] and the review article [21]. There are mainly two variants of the GLSFEMs; the stabilized and the direct versions.

Stabilized finite elements method [57]: The standard (continuous) Galerkin FEM for the problem (3.34) reads: find $u_h \in U_h \subset U$ (U: solution space) such that

$$a(u_h, v_h) + (r(u_h), v_h)_{L^2(\Omega)} = (f, v_h)_{L^2(\Omega)} , \quad \forall v_h \in V_h \subset V \qquad (3.35)$$

where $a(u,v) = (\varepsilon \nabla u + \beta \cdot \nabla u + \alpha u, v)_{L^2(\Omega)}$ is the standard bilinear form to the linear part of (3.34). The general stabilized FEMs formulation reads as: for all $v_h \in V_h \subset V$, find $u_h \in U_h \subset U$ such that

$$a(u_h, v_h) + (r(u_h), v_h)_{L^2(\Omega)} + \sum_K \tau_K S_K(\mathbf{u_h}, \mathbf{v_h}) = (f, v_h)_{L^2(\Omega)} \qquad (3.36)$$

where the stabilization parameter is defined on each element K as [52]

$$\tau_K = \frac{1}{\frac{4\varepsilon}{h^2} + \frac{2|\beta|}{h} + |\alpha|}.$$

One possible approach to construct GLSFEM approximation is to use the least-squares term $\tilde{J}_K(u,v)$ as the stabilization term S_K in (3.36), i.e.: for all $v_h \in V_h$, find $u_h \in U_h$ such that

$$a(u_h, v_h) + (r(u_h), v_h)_{L^2(\Omega)} + \sum_K \tau_K \tilde{J}_K(u_h, v_h) = (f, v_h)_{L^2(\Omega)}. \qquad (3.37)$$

Note that the stabilized continuous FEM, streamline upwind Petrov-Galerkin (SUPG) method is obtained by setting

$$S_K(u_h, v_h) = (\mathscr{L}u_h + r(u_h) - f, \beta \cdot \nabla v_h)_{L^2(K)}$$

with different choices of the parameter τ_K.

The direct variant of GLSFEM: Another approach to construct the GLSFEM is to discretize just the least-squares term $\tilde{J}_{\Omega}(u,v)$ by solving his problem as: : for all $v_h \in V_h \subset H^2(\Omega) \cap V$, find $u_h \in U_h \subset H^2(\Omega) \cap U$ such that $\tilde{J}_{\Omega}(u_h, v_h) = 0$, i.e.,

$$\int_{\Omega} (\mathscr{L}u_h + r(u_h))(\mathscr{L}v_h + r'(u_h)v_h)dx = \int_{\Omega} f(\mathscr{L}v_h + r'(u_h)v_h)dx$$

which is not only a fourth order problem but also the solution and trial subspaces U_h and V_h need to be continuously differentiable functions. This makes the construction bases functions complicated, whereas the standard finite element spaces and the assembly of the stiffness matrix can no longer be used. The condition number of the stiffness is order of $\mathscr{O}(h^{-4})$ instead of being order of $\mathscr{O}(h^{-2})$ for the standard Galerkin FEM. Hence, this approach is impractical. As a remedy for this, the problem (3.34) is converted into a first-order system [54, 21]:

$$p - \nabla u = 0, \qquad \text{in } \Omega,$$
$$-\varepsilon \nabla \cdot p + \beta \cdot \nabla u + \alpha u + r(u) = f, \qquad \text{in } \Omega,$$
$$u = g, \qquad \text{on } \Gamma.$$

Then, we define now the least-square functional for $z = (p, u)^T$ as

$$J(z) := \frac{1}{2} \|p - \nabla u\|_{L^2(\Omega)}^2 + \frac{1}{2} \| - \varepsilon \nabla \cdot p + \beta \cdot \nabla u + \alpha u + r(u) - f\|_{L^2(\Omega)}^2.$$

A minimizer of $J(z)$ is obtained by the identity

$$\lim_{t \to 0} \frac{d}{dt} J(z + tv) = 0, \quad \forall v$$

which yields a least-squares term of order two. Using this approach, we solve the resulting least-squares term, which is a second-order problem now, using continuous or discontinuous finite element test and trial spaces $S_h \subset H^1(\Omega, div) \times U$ ($S_h \subset H^1(\Omega, div) \times H^1(\Omega)$) and $T_h \subset H^1(\Omega, div) \times V$ ($T_h \subset H^1(\Omega, div) \times H^1(\Omega)$), respectively. The condition number of the stiffness matrix is retained as $\mathscr{O}(h^{-2})$ as in the standard Galerkin method [21]. For advection dominated problems, the resulting linear systems of equations are solved usually with a preconditioned conjugate gradient method due to large condition numbers, as reported in [64] for a GLSFEM solution of singularly perturbed advection-diffusion problems.

In order to compare the GLSFEM with the dGFEM and dGAFEM, we consider the linear problem [97]

$$-\varepsilon \Delta u + \beta \cdot \nabla u + \alpha u = f \quad \text{in } (0,1)^2 \tag{3.38}$$

with $\varepsilon = 10^{-6}$, $\beta = (2,3)^T$ and $\alpha = 1$. The load function f and Dirichlet boundary conditions are chosen so that the exact solution is

$$u(x_1, x_2) = \frac{\pi}{2} \arctan\left(\frac{1}{\sqrt{\varepsilon}}(-0.5x_1 + x_2 - 0.25)\right).$$

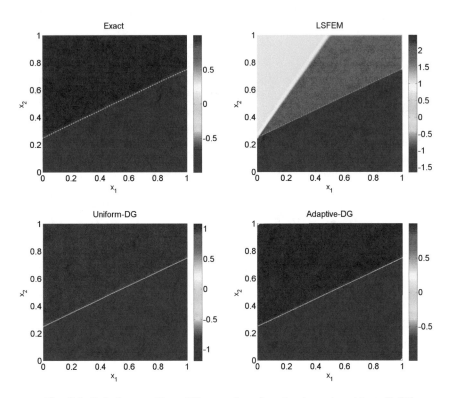

Fig. 3.3: Solution profiles of linear advection dominated problem (3.38)

Fig.3.3 shows that GLSFEM is not stable for the non-linear model (3.38), and even a uniform dG scheme is capable of solving the problem. We can see from Fig.3.4 the global L^2-errors and the result that the solutions obtained by dGAFEM produces more accurate profiles than the dGFEM for the advection dominated problem (3.38).

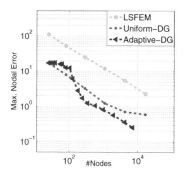

Fig. 3.4: Comparison of the GLSFEM and dGFEM for linear advection dominated problem (3.38)

3.5 Numerical Examples

In this section, we give several numerical examples demonstrating the effectiveness and accuracy of the dGAFEM for stationary advection dominated semi-linear ADR equations of the form (3.1).

3.5.1 Example with Polynomial Type Non-Linearity

Our first example is taken from [15] with Dirichlet boundary condition on $\Omega = (0,1)^2$ with $\varepsilon = 10^{-6}$, $\beta = \frac{1}{\sqrt{5}}(1,2)^T$, $\alpha = 1$ and $r(u) = u^2$. The source function f and Dirichlet boundary condition are chosen so that

$$u(x_1, x_2) = \frac{1}{2}\left(1 - \tanh\frac{2x_1 - x_2 - 0.25}{\sqrt{5\varepsilon}}\right)$$

is the exact solution. The problem is characterized by an internal layer of thickness $\mathcal{O}(\sqrt{\varepsilon}\,|\ln\varepsilon\,|)$ around $2x_1 - x_2 = \frac{1}{4}$.

The mesh is locally refined by dGAFEM around the interior layer (Fig.3.5) and the spurious solutions are damped out in Fig.3.6, similar to the results as in [15] using SUPG-SC, in [98] with SIPG-SC. On adaptively and uniformly refined meshes, from the Fig.3.7, it can be clearly seen that the adaptive meshes reduce the substantial computing time. On uniform meshes, the SIPG is slightly more accurate as shown in [98] than the SUPG-SC in [15]. The error reduction by increasing degree of the polynomials is remarkable on finer adaptive meshes (Fig.3.7, right).

In Table 3.1, we give the results using the solution technique in Section 3.3.1 for the *BiCGStab* with the stopping criterion as $\|r_i\|_2/\|r_0\|_2 \leq tol$ for $tol = 10^{-7}$ (r_i is the residual of the corresponding system at the ith iteration) applied to the

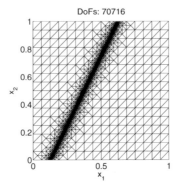

Fig. 3.5: Example 3.5.1; Adaptive mesh

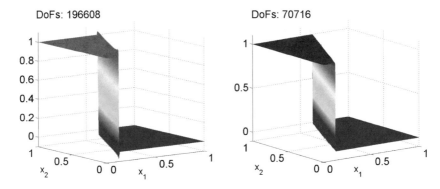

Fig. 3.6: Example 3.5.1: Uniform (left) and adaptive (right) solutions, quadratic elements

unpermuted system and Schur complement system with and without precondition-ing on the finest levels of uniformly (4th refinement level with DoFs 196608 and 32768 triangular elements) and adaptively (17th refinement level with DoFs 70716 and 11786 triangular elements) refined meshes. As a preconditioner, the incomplete LU factorization of the Schur complement matrix S (ILU(S)) is used for the linear system with the coefficient matrix S. The linear systems with the coefficient matrix A are solved directly. Table 3.1 shows that solving the problem via the block LU factorization using the Schur complement system with the preconditioner ILU(S) is the fastest.

The time for applying the permutation to obtain the reordered matrix and the permutation matrix P takes 9.9 seconds, whereas, it takes 0.13 seconds to form the Schur complement matrix S and 0.04 seconds to compute the ILU(S) on a PC with Intel Core-i7 processor and 8GB RAM using the 64-bit version of Matlab-R2010a. We note that since the Jacobian matrix does not change during the non-linear itera-

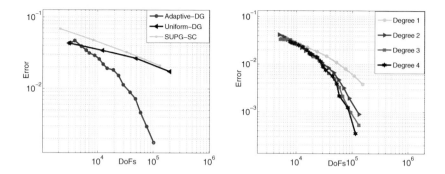

Fig. 3.7: Example 3.5.1: Global errors: comparison of the methods by quadratic elements (left), adaptive dG for polynomial degrees 1-4 (right)

Table 3.1: Example 3.5.1: Average number of Newton iterations, average number of *BiCGStab* iterations, total computation time in seconds corresponding to the uniformly refined (adaptively refined) mesh

Linear Solver	# Newton	# BiCGStab	Time
BiCGStab w/o prec. (Unpermuted)	10.8 (10.5)	818 (757.5)	1389.3 (773.3)
BiCGStab w/ prec. M_1 (Permuted)	10.3 (10.3)	1.5 (3)	423.1 (374.2)
BiCGStab w/ prec. M_2 (Permuted)	10.3 (10.3)	1.5 (3)	416.8 (375.9)
Block LU + (BiCGStab w/o prec.)	10.3 (10.9)	247.5 (315.5)	270.9 (310.3)
Block LU + (BiCGStab w/ prec. ilu(S))	10.3 (10.9)	19 (28.5)	140.9 (114.7)

tions, the permutation, the Schur complement matrix and ILU(S) is computed only once for each run.

In all of the following results and figures, the Jacobian matrix J is scaled by a left Jacobi preconditioner before reordering to obtain a well conditioned matrix. The reordering procedure is applied to the scaled Jacobian matrix. Reordering times, which are included in the total computation time, for the uniform and adaptive refinements are 102.1 seconds and 41.4 seconds, respectively.

Fig.3.10 shows the condition numbers of the Jacobian matrices J of the original system, S and A of the block LU factorized system on the uniformly and adaptively refined meshes. The condition numbers of the coefficient matrix A are almost constant for uniform refinement by different orders of dG discretizations and around 10, whereas the condition number of S is lower than of the Jacobian matrix J. This is due to the clustering of nonzero elements around the diagonal (Fig.3.11) due to the matrix reordering. For adaptive refinement, Fig.3.10, right, we observe the same behavior, whereas the condition numbers are larger of order one than for the uniform refinement. For comparison, we provide results by using BiCGStab with two block preconditioners. The preconditioning matrices M_1 and M_2 for the permuted full systems are given as

$$M_1 = \begin{pmatrix} A & 0 \\ C^T & S \end{pmatrix}, \qquad M_2 = \begin{pmatrix} A & B \\ 0 & S \end{pmatrix}.$$

The total number of iterations and time for different algorithms are given in Table 3.1. Our proposed method where we compute the block LU factorization of the partitioned matrix and solve the system involving the Schur complement iteratively via preconditioned BiCGStab is the best in terms of the total time compared to other methods for both uniform and adaptive refinement. In Fig.3.8 and Fig.3.9, we present the total time and the average number of linear solver iterations, respectively, for uniform and adaptive refinements as the problem size has been increased. We observe that the proposed preconditioned linear solver has been the best in terms of time with a reasonable number of iterations for different problem sizes regardless of refinement type.

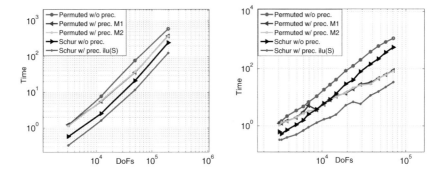

Fig. 3.8: Example 3.5.1: Computation time vs. DoFs: Uniform refinement (left) and adaptive refinement (right)

3.5.2 Example with Monod Type Non-Linearity

We consider the problem in [15] of type (3.1) on $\Omega = (0,1)^2$, and having a Monod type non-linearity $r(u) = -u/(1+u)$ and homogeneous source function. The advection field and the diffusion coefficient are given as $\beta(x_1,x_2) = (-x_2,x_1)^T$ and $\varepsilon = 10^{-6}$, respectively. The Dirichlet boundary condition is prescribed as

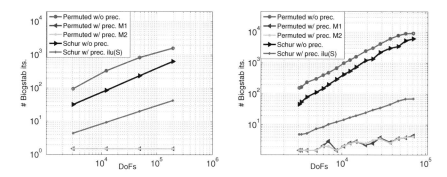

Fig. 3.9: Example 3.5.1: # Average *BiCGStab* iterations vs. DoFs: Uniform refinement (left) and adaptive refinement (right)

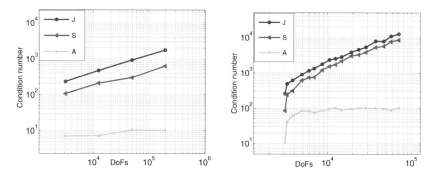

Fig. 3.10: Example 3.5.1: Condition number of the matrices J (unpermuted matrix), S (Schur complement matrix) and A (left top block of permuted matrix): Uniform refinement (left) and adaptive refinement (right)

$$u(x_1, x_2) = \begin{cases} 1 & \text{for } 1/3 \leq x_1 \leq 2/3, \ x_2 = 0, \\ 0 & \text{for } x_1 < 1/3, x_1 > 2/3, \ x_2 = 0, \\ 0 & \text{for } x_1 = 1 \text{ or } x_2 = 1. \end{cases}$$

On the left boundary ($x_1 = 0, 0 \leq x_2 \leq 1$), no-flow condition is assumed.

There are both internal and boundary layers on the mesh (Fig.3.12, left), around them oscillations occur. Fig.3.12, right, shows that by dGAFEM, the oscillations almost disappear, similar to the results in [15] for the SUPG-SC and in [98] for SIPG-SC. Fig.3.12, left, shows that the adaptive process leads to correctly refined meshes. Moreover, by increasing polynomial degree ($k = 4$), the oscillations are completely eliminated on the outflow boundary (Fig.3.12, bottom) and the sharp

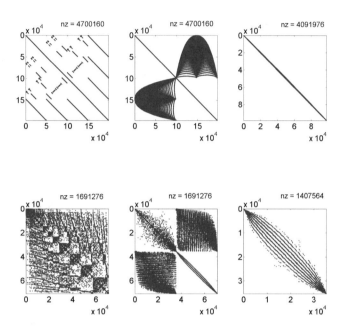

Fig. 3.11: Example 3.5.1: Sparsity patterns of the unpermuted (left), permuted (middle) and the Schur complement (right) matrices at the final refinement levels: Uniform refinement (top) with DoFs 196608 and adaptive refinement (bottom) with DoFs 70716

front is preserved. This is not the case for SUPG-SC [15] and SIPG-SC [98], where still small oscillations are present.

As in the case of polynomial non-linearity, Example 3.5.1, the block LU factorized system solved by BiCGStab with the preconditioner ILU(S) is the most efficient solver, with an average number of 7 Newton iterations. The computing times for the uniform refinement was 20.6 seconds, and 30.5 for the adaptive refinement.

3.5.3 Coupled Example with Arrhenius Type Non-Linearity

The next example is the non-linear reaction for a two-component system in [87]:

$$-\nabla \cdot (\varepsilon \nabla u_1) + \beta \cdot \nabla u_1 - 100k_0 u_2 e^{\frac{-E}{R u_1}} = 0,$$
$$-\nabla \cdot (\varepsilon \nabla u_2) + \beta \cdot \nabla u_2 + k_0 u_2 e^{\frac{-E}{R u_1}} = 0,$$

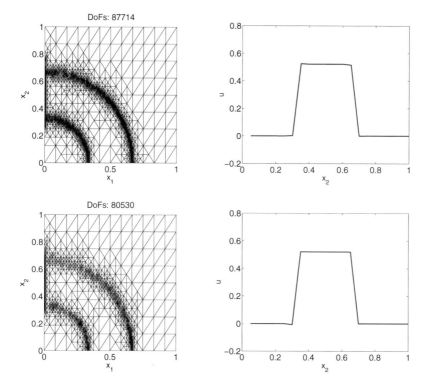

Fig. 3.12: Example 3.5.2: Adaptive meshes (left) and the cross-section plots (right) of the solutions at the left outflow boundary by quadratic elements (top) and quartic elements (bottom)

on $\Omega = (0,1)^2$ with the advection field $\beta = (1 - x_2^2, 0)^T$, the diffusion constant $\varepsilon = 10^{-6}$, the reaction rate coefficient $k_0 = 3 \times 10^8$ and the quotient of the activation energy to the gas constant $E/R = 10^4$. The unknowns u_1 and u_2 represent the temperature of the system and the concentration of the reactant, respectively.

There are oscillations around the layers, even small, for the uniform refinement (Fig.3.13, left) as for SIPG-SC in [98]. On the other hand, these oscillations are completely dumped out by dGAFEM with almost half of the DoFs used in the uniform refinement (Fig.3.13, right).

The block LU factorization based algorithm with the preconditioner ILU(S) requires 10.5 seconds for the uniform and 24.4 seconds for the adaptive refinements. Matrix reordering and permutation took 2.44 seconds for the uniform and 2.17 seconds for adaptive refinements, respectively.

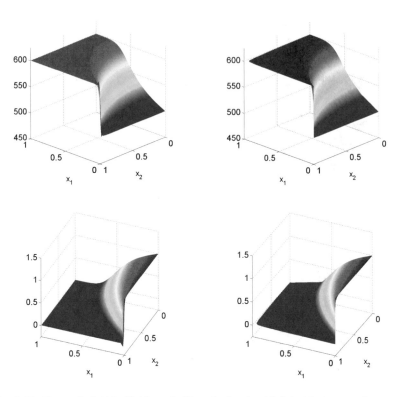

Fig. 3.13: Example 3.5.3: Uniform(left) and adaptive(right) solutions to the temper-
ature(top) and reactant(bottom), quadratic elements with DoFs 12288 for uniform
refinement and with DoFs 6168 for adaptive refinement

3.5.4 Coupled Example with Polynomial Type Non-Linearity

Our final problem is the modification of the non-stationary transport problem, Example 2, in [17]. The problem is stated as the following:

$$\alpha u_1 - \nabla \cdot (\varepsilon \nabla u_1) + \beta \cdot \nabla u_1 + 50 u_1^2 u_2^2 = 0,$$
$$\alpha u_2 - \nabla \cdot (\varepsilon \nabla u_2) + \beta \cdot \nabla u_2 + +50 u_1^2 u_2^2 = 0,$$

on the rectangular domain $\Omega = (0,1) \times (0,2)$ with the advection field $\beta = (0,-1)^T$,
the diffusion constant $\varepsilon = 10^{-10}$ and linear reaction constant $\alpha = 0.1$. On the left,
right and lower parts of the boundary of the domain, Neumann boundary conditions
are prescribed. On the remaining part of the boundary, Dirichlet boundary conditions are chosen as

$$u_1(\mathbf{x}) = \begin{cases} 8(x_1 - 0.375) & \text{for } 0.375 < x_1 \leq 0.5, \\ -8(x_1 - 0.625) & \text{for } 0.5 < x_1 \leq 0.625, \\ 0 & \text{otherwise} \end{cases}$$

$$u_2(\mathbf{x}) = \begin{cases} 8(x_1 - 0.125) & \text{for } 0.125 \leq x_1 \leq 0.25, \\ -8(x_1 - 0.375) & \text{for } 0.25 < x_1 \leq 0.375, \\ 8(x_1 - 0.625) & \text{for } 0.625 \leq x_1 \leq 0.75, \\ -8(x_1 - 0.875) & \text{for } 0.75 < x_1 \leq 0.875, \\ 0 & \text{otherwise.} \end{cases}$$

There is a boundary layer on the outflow boundary, Fig.3.15. Fig.3.14 shows that oscillations are almost damped using dGAFEM approximations, similar to those results in [17] using SUPG-SC. It can be seen from Fig.3.15 that the mesh is correctly refined by dGAFEM near the boundary layer.

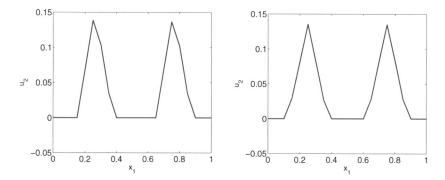

Fig. 3.14: Example 3.5.4: Uniformly (left) and adaptively (right) obtained cross-section plots on the outflow boundary for the component u_2, quartic elements with DoFs 61440 for uniform refinement and with DoFs 33690 for adaptive refinement

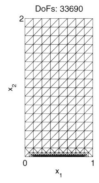

Fig. 3.15: Example 3.5.4: Adaptive mesh, quartic elements with DoFs 33690

Chapter 4
Parabolic Problems with Space-Time Adaptivity

Application of adaptive dG methods and a posteriori error estimates to problems in geoscience are reviewed recently in [33]. Most of the applications of dG methods in geoscience concern reactive transport with advection [13, 62, 84] and strong permeability contrasts such as layered reservoirs [90] or vanishing and varying diffusivity posing challenges in computations [72]. The permeability in heterogeneous porous and fractured media varies over orders of magnitude in space, which results in highly variable flow field, where the local transport is dominated by advection or diffusion [85]. Accurate and efficient numerical solution of ADR equations to predict macroscopic mixing, anomalous transport of solutes and contaminants for a wide range of parameters like permeability and Péclet numbers, different flow velocities and reaction rates and reaction rates are challenging problems [85]. In order to resolve the complex flow patterns accurately, higher order time stepping methods like exponential time stepping methods are used [85].

The solution of the evolution problems modeled by advection dominated non-stationary ADR equations has a number of challenges. On one hand, the solution around the interior/boundary layers due to large advective terms has to be resolved. On the other hand, the nature of non-stationary models leads to resolution of spatial layers to be more critical since the location of the layers may vary with time, and it is possible that temporal layers also occur in addition to the spatial layers. The usual approach in such cases is the use of adaptive algorithms to resolve the solution in an accurate and efficient way around the regions where the solution obeys large gradients. The key tool in an adaptive algorithm, thus, is a way of locating the so-called regions, which is usually and naturally based on a posteriori error estimation. Besides, even for linear evolution problems, the results for a posteriori error estimation in evolution problems are limited. The analysis of existence studies are mostly based on energy techniques. We refer to the studies in [1, 9, 20, 30, 71] and references therein.

Being a native approach in evolution problems, a posteriori error estimation based on energy techniques compares the continuous and the discrete solution directly. However, the driven a posteriori error bounds, then, are optimal order in $L^2(H^1)$-type norms, but sub-optimal order in $L^\infty(L^2)$-type norms. Further, the nu-

© Springer International Publishing Switzerland 2016

M. Uzunca, *Adaptive Discontinuous Galerkin Methods for Non-linear Reactive Flows*,
Lecture Notes in Geosystems Mathematics and Computing,
DOI 10.1007/978-3-319-30130-3_4

merical solution does not satisfy the point-wise structural conditions of the true solution. For this reason, in this chapter, we develop for the model (1.1) a space-time adaptive algorithm using SIPG in space and backward Euler in time in order to achieve the optimal orders in both $L^2(H^1)$ and $L^\infty(L^2)$-type norms, and to insure that the numerical solution satisfies the point-wise structural conditions of the true solution.

4.1 Preliminaries and Model Equation

Let $\Omega \subset \mathbb{R}^2$ be a bounded, open and convex domain with boundary $\partial\Omega$. For a Banach space X, define the spaces $L^p(0,T;X)$:

$$\|v\|_{L^p(0,T;X)} = \left(\int_0^T \|v(t)\|_X^p dt \right)^{1/p} < \infty, \qquad \text{for } 1 \le p < +\infty,$$

$$\|v\|_{L^\infty(0,T;X)} = \operatorname*{esssup}_{0 \le t \le T} \|v(t)\|_X < \infty, \qquad \text{for } p = +\infty.$$

We also introduce the space

$$H^1(0,T;X) = \{v \in L^2(0,T;X) | \ v_t \in L^2(0,T;X)\}.$$

We denote by $C(0,T;X)$ and $C^{0,1}(0,T;X)$ the spaces of continuous and Lipschitz-continuous functions $v : [0,T] \mapsto X$, respectively, equipped with the norms

$$\|v\|_{C(0,T;X)} = \max_{0 \le t \le T} \|v(t)\|_X < \infty,$$

$$\|v\|_{C^{0,1}(0,T;X)} = \max \left\{ \|v\|_{C(0,T;X)}, \|v_t\|_{C(0,T;X)} \right\} < \infty.$$

The model equation we consider in this chapter is the non-stationary system of semi-linear advection-diffusion-reaction equations of type (1.1) for an indexed set $i = 1,2,\ldots,m$ given by

$$\frac{\partial u_i}{\partial t} - \nabla \cdot (\varepsilon_i \nabla u_i) + \beta_i \cdot \nabla u_i + r_i(\mathbf{u}) = f_i \qquad (4.1)$$

in $\Omega \times (0,T]$ for the vector of unknowns $\mathbf{u} = \mathbf{u}(x,t) = (u_1,u_2,\ldots,u_m)^T$ with appropriate boundary and initial conditions, and with $u_j = u_j(x,t)$, $j = 1,2,\ldots,m$. We assume that the source functions $f_i \in C(0,T;L^2(\Omega))$, and the velocity fields $\beta_i \in C\left(0,T;W^{1,\infty}(\Omega)\right)^2$ are either given or computed. In the models for flow in heterogeneous media, the symmetric dispersion tensor ε_i is of the form

$$\varepsilon_i = \begin{bmatrix} D_i^1 & 0 \\ 0 & D_i^2 \end{bmatrix}$$

with $0 < D_i^1, D_i^2 \ll 1$. Moreover, we assume that the non-linear reaction terms are bounded, locally Lipschitz continuous and monotone; they satisfy for any $s, s_1, s_2 \geq 0$, $s, s_1, s_2 \in \mathbb{R}$ the following conditions:

$$|r_i(s)| \leq C_S, \quad C_S > 0, \ s \in [-S, S], \tag{4.2a}$$

$$\|r_i(s_1) - r_i(s_2)\|_{L^2(\Omega)} \leq L(S)\|s_1 - s_2\|_{L^2(\Omega)}, \quad L > 0, \tag{4.2b}$$

$$r_i \in C^1(\mathbb{R}_0^+), \quad r_i(0) = 0, \quad r_i'(s) \geq 0. \tag{4.2c}$$

We further assume that there are $\alpha_0, c_* \geq 0$ satisfying for $i = 1, 2, \ldots, m$,

$$\| - \nabla \cdot \beta_i(x, t)\|_{C(0, T; L^\infty(\Omega))} \leq c_* \alpha_0. \tag{4.3}$$

In the sequel, for simplicity, we just consider a single equation of the system (4.1) (m=1) to apply the SIPG discretization in space with a homogeneous dispersion tensor leading to a simple diffusivity constant $0 < \varepsilon \ll 1$. We further take into account the homogeneous Dirichlet boundary condition on the whole boundary to simplify the a posteriori error constructions. The a posteriori error analysis can be extended to the problems with heterogeneous dispersion tensor and inhomogeneous Dirichlet or Neumann boundary condition in a standard way.

Thus, we consider the semi-linear system

$$\frac{\partial u}{\partial t} - \varepsilon \Delta u + \beta \cdot \nabla u + r(u) = f \qquad \text{in } \Omega \times (0, T], \tag{4.4a}$$

$$u(\cdot, t) = 0 \qquad \text{on } \partial\Omega \times (0, T], \tag{4.4b}$$

$$u(\cdot, 0) = u_0 \qquad \text{in } \Omega. \tag{4.4c}$$

Then, the standard weak formulation of (4.4) reads as: for any $v \in H_0^1(\Omega)$, find $u \in L^2(0, T; H_0^1(\Omega)) \cap H^1(0, T; L^2(\Omega))$ satisfying, for all $t \in (0, T]$,

$$\int_\Omega \frac{\partial u}{\partial t} v \, dx + a(t; u, v) + b(t; u, v) = l(v), \tag{4.5}$$

$$a(t; u, v) = \int_\Omega (\varepsilon \nabla u \cdot \nabla v + \beta \cdot \nabla u v) dx, \tag{4.6a}$$

$$b(t; u, v) = \int_\Omega r(u) v \, dx, \tag{4.6b}$$

$$l(v) = \int_\Omega f v \, dx, \tag{4.6c}$$

which have a unique solution in the space $C(0, T; L^2(\Omega))$ under the given regularity assumptions and the conditions (4.2a)-(4.2c). Further, using the definitions above, one can easily show that the bilinear form $a(t; u, v)$ is coercive and continuous on the space $H_0^1(\Omega)$ such that

$$a(t;v,v) \geq |||v|||^2 \qquad\qquad\qquad\qquad \forall v \in H_0^1(\Omega), \qquad (4.7)$$

$$a(t;u,v) \leq ||u||_{dG}|||v||| \qquad \forall v \in H_0^1(\Omega), \ \forall u \in H_0^1(\Omega) + V_h. \qquad (4.8)$$

4.2 Semi-Discrete and Fully Discrete Formulations

In this section, we introduce the semi-discrete and fully-discrete formulations of the model (4.4). Most of the notations here are analogue to those in Chapter 3, but now time-dependent.

4.2.1 Semi-Discrete Formulation

Let the mesh $\xi = \{K\}$ be a family of shape regular elements (triangles). We set the mesh-dependent finite dimensional solution and test function space by

$$V_h = V_h(\xi) = \{v \in L^2(\Omega) : v|_K \in \mathbb{P}_k(K), \forall K \in \xi\} \not\subset H_0^1(\Omega).$$

For a given $t \in [0,T]$, we restate the sets of inflow and outflow edges by

$$\Gamma_t^- = \{x \in \partial\Omega : \beta(x,t) \cdot \mathbf{n}(x) < 0\}, \qquad \Gamma_t^+ = \partial\Omega \setminus \Gamma_t^-,$$

$$\partial K_t^- = \{x \in \partial K : \beta(x,t) \cdot \mathbf{n}_K(x) < 0\}, \qquad \partial K_t^+ = \partial K \setminus \partial K_t^-.$$

Moreover, we denote by Γ_h^0 and Γ_h^∂ the set of interior and boundary edges, respectively, so that the union set $\Gamma_h = \Gamma_h^0 \cup \Gamma_h^\partial$ forms the skeleton of the mesh. Then, utilizing the SIPG construction in Chapter 3, the semi-discrete formulation of (4.4) reads as: for $t = 0$ set $u_h(0) \in V_h(\xi)$ as the projection (orthogonal L^2-projection) of u_0 onto $V_h(\xi)$, and for each $t \in (0,T]$, for all $v_h \in V_h(\xi)$, find $u_h \in C^{0,1}(0,T;V_h(\xi))$ such that

$$\int_\Omega \frac{\partial u_h}{\partial t} v_h dx + a_h(t;u_h,v_h) + K_h(u_h,v_h) + b_h(t;u_h,v_h) = l_h(v_h), \qquad (4.9)$$

where the forms are given by

$$a_h(t;u_h,v_h) = \sum_{K\in\xi}\int_K \varepsilon\nabla u_h \cdot \nabla v_h dx + \sum_{K\in\xi}\int_K \beta\cdot\nabla u_h v_h dx$$

$$+ \sum_{e\in\Gamma_h}\frac{\sigma\varepsilon}{h_e}\int_e [u_h]\cdot[v_h]ds - \sum_{K\in\xi}\int_{\partial K_t^-\cap\Gamma_t^-}\beta\cdot\mathbf{n}_K u_h v_h ds$$

$$+ \sum_{K\in\xi}\int_{\partial K_t^-\backslash\partial\Omega}\beta\cdot\mathbf{n}_K(u_h^{out}-u_h)v_h ds, \tag{4.10a}$$

$$K_h(u_h,v_h) = -\sum_{e\in\Gamma_h}\int_e (\{\varepsilon\nabla v_h\}\cdot[u_h]+\{\varepsilon\nabla u_h\}\cdot[v_h])ds, \tag{4.10b}$$

$$b_h(t;u_h,v_h) = \sum_{K\in\xi}\int_K r(u_h)v_h dx, \tag{4.10c}$$

$$l_h(v_h) = \sum_{K\in\xi}\int_K f_h v_h dx. \tag{4.10d}$$

Upon integration by parts on the advective term, bilinear form (4.10a) will be

$$a_h(t;u_h,v_h) = \sum_{K\in\xi}\int_K \varepsilon\nabla u_h\cdot\nabla v_h dx - \sum_{K\in\xi}\int_K (\beta u_h\cdot\nabla v_h+\nabla\cdot\beta u_h v_h)dx$$

$$+ \sum_{e\in\Gamma_h}\frac{\sigma\varepsilon}{h_e}\int_e [u_h]\cdot[v_h]ds + \sum_{K\in\xi}\int_{\partial K_t^+\cap\Gamma_t^+}\beta\cdot\mathbf{n}_K u_h v_h ds$$

$$+ \sum_{K\in\xi}\int_{\partial K_t^+\backslash\partial\Omega}\beta\cdot\mathbf{n}_K u_h(v_h-v_h^{out})ds. \tag{4.11a}$$

Note that the bilinear form $a_h(t;u,v)$ in (4.10a) is well-defined for the functions $u,v\in H_0^1(\Omega)$

$$a_h(t;u,v) = \int_\Omega (\varepsilon\nabla u\cdot\nabla v+\beta\cdot\nabla uv)dx.$$

Thus, the continuous weak formulation (4.5) can be rewritten for any $t\in(0,T]$ as

$$\int_\Omega \frac{\partial u}{\partial t}v dx + a_h(t;u,v)+b(t;u,v) = l(v), \qquad \forall v\in H_0^1(\Omega). \tag{4.12}$$

Proposition 4.1 (Existence of Unique Solution). *The SIPG semi-discrete system (4.9) has a unique solution in* $C^{0,1}(0,T;V_h(\xi))$.

Proof. Using the matrix-vector notations and construction introduced in Section 3.1.1, we obtain the following system of non-linear equations

$$M\mathbf{U}_t + S\mathbf{U} = \mathbf{L} - \mathbf{b}(\mathbf{U}), \tag{4.13}$$

where $\mathbf{U}$ is the vector of unknown coefficients, the matrix S is the stiffness matrix corresponding to bilinear form $a_h(t;u_h,v_h)+K_h(u_h,v_h)$, vector functions $\mathbf{b}(\mathbf{U})$, $\mathbf{L}$ correspond to the non-linear forms $b_h(t;u_h,v_h)$ and the linear form $l_h(v_h)$, respectively. The matrix M is the usual mass matrix, a symmetric positive definite matrix.

Altogether, the dG space discretization results an ODE system (4.13) with the invertible mass matrix M and the right-hand side is Lipschitz with respect $\mathbf{U}$, which implies the ODE system has a unique solution.

4.2.2 Fully Discrete Formulation

For discretization in time, we use the backward Euler method which is an unconditionally stable integrator for stiff ODEs. We consider a subdivision of $[0, T]$ into n time intervals $I_k = (t^{k-1}, t^k]$ of length Δt_k, $k = 1, 2, \ldots, n$. Set $t^0 = 0$ and for $n \geq 1$, $t^k = \Delta t_1 + \Delta t_2 + \cdots + \Delta t_k$. Denote by ξ^0 an initial triangulation and by ξ^k the mesh associated to the kth time step for $k > 0$, which is obtained from ξ^{k-1} possibly by locally refining/coarsening. Moreover, we assign the finite element space $V_h^k = V_h(\xi^k)$ to each mesh ξ^k. Then, backward Euler in time the fully discrete formulation of (4.9) reads as: for $t = 0$ set $u_h^0 \in V_h^0$ as the projection (orthogonal L^2-projection) of u_0 onto V_h^0 and for $k = 1, 2, \ldots, n$, find $u_h^k \in V_h^k$ satisfying, for all $v_h^k \in V_h^k$,

$$\int_\Omega \frac{u_h^k - u_h^{k-1}}{\Delta t_k} v_h^k dx + a_h(t^k; u_h^k, v_h^k) + K_h(u_h^k, v_h^k)$$
$$+ b_h(t^k; u_h^k, v_h^k) = \int_\Omega f^k v_h^k dx. \tag{4.14}$$

Remark 4.1. A single system of (4.14) is considered as a non-stationary semi-linear ADR equation of type (3.1) with the linear reaction coefficient $\alpha = 1/\Delta t_k > 0$ and with a modified right-hand side. Hence, it is a direct consequence not only by the existence of unique solution result given for the semi-discrete system (4.9) but also by the ones given for stationary problems that for each $k = 1, 2, \ldots, n$, the full-discrete system (4.14) has a unique solution.

4.3 Space-Time Adaptivity for Non-Stationary Problems

Standard energy techniques for evolution problems are not very suitable for space-time adaptive methods. They, firstly, do not result in optimal convergence rates in $L^\infty(L^2)$-type norms. Moreover, the numerical solution does not satisfy the pointwise structural conditions of the true solution. The key reason of these issues is that in standard energy techniques, the numerical solution is compared with the true solution directly. In this chapter, to construct the space-time adaptive algorithm, we derive a posteriori error estimates for semi-linear problems of the form (1.1) by comparing the numerical solution, in contrast to the standard energy techniques, with an auxiliary solution but not with the true solution. We use the *elliptic reconstruction* technique in [67] which allows us to utilize the ready a posteriori error estimates for elliptic models derived in Chapter 3 to control the spatial error. The idea of the elliptic reconstruction technique is to construct an auxiliary solution whose difference from the numerical solution can be estimated by a known (elliptic) a posteriori estimate, and the constructed auxiliary solution satisfies a variant of the given problem with a right-hand side which can be controlled in an optimal way. Furthermore, the obtained error estimates are of optimal order in both $L^2(H^1)$ and $L^\infty(L^2)$-type norms. In [26], a posteriori error estimates in the $L^\infty(L^2) + L^2(H^1)$-type norm are derived for linear ADR equations using backward Euler in time and dG in space utilizing the elliptic reconstruction technique. In this section, we extend the study in [26] to the advection dominated semi-linear ADR equations.

To motivate the readers, we start with the simple evolution problem

$$
\begin{aligned}
u'(t) + A(u(t)) &= f && \text{in } \Omega \times (0, T], \\
u(\cdot, t) &= 0 && \text{on } \partial\Omega \times (0, T], \\
u(\cdot, 0) &= u_0 && \text{in } \Omega,
\end{aligned}
$$

leading to the weak formulations

$$
\langle u'(t), v \rangle + a(u(t), v) = \langle f, v \rangle, \quad \forall v \in V, \text{ a.e. } t \in (0, T], \tag{4.15}
$$
$$
\langle U'(t), v \rangle + a(U(t), v) = \langle f, v \rangle, \quad \forall v \in V_h, \text{ a.e. } t \in (0, T] \tag{4.16}
$$

with the exact and numerical solutions u and U, respectively, in the appropriate solution spaces. Then, the direct approach adaptive algorithms aim to derive an energy norm based on a posteriori error indicator η to estimate the error between u and U by a suitable energy norm $\|\cdot\|_E$, i.e. find η satisfying

$$
\|u - U\|_E \leq \eta(U).
$$

The native way of deriving such a residual-based a posteriori error indicator is to estimate, using the Galerkin orthogonality property, the residual function R under the given energy norm, and satisfying the variational formulation

$$
\begin{aligned}
\langle R_U, v \rangle &= \int_0^T \left(\langle f, v \rangle - \langle U'(t), v \rangle - a(U(t), v) \right) dt \\
&= \int_0^T \left(\langle f, v - I_h v \rangle - \langle U'(t), v - I_h v \rangle - a(U(t), v - I_h v) \right) dt.
\end{aligned}
$$

The above identity leads to the error equation

$$
\frac{1}{2} \|u(T) - U(T)\|_{L^2(\Omega)}^2 + \int_0^T a(u - U, u - U) dt = \frac{1}{2} \|u(0) - U(0)\|_{L^2(\Omega)}^2 + \langle R_U, v \rangle
$$

where it is well-known that the integral term is sub-optimal in $L^\infty(L^2)$-type norms. On the other hand, the residual R_U poses a number of challenges:

- It is possible that $\langle R_U, v \rangle$ leads to a singular vector,
- The numerical solution U does not satisfy point-wise structural conditions of the true solution u.

To handle the above drawbacks a continuous auxiliary solution $\tilde{U}$ is constructed which is easily computable from the numerical solution U through a suitable operator, and it satisfies the following conditions:

- The difference $U - \tilde{U}$ is computable and can be estimated by a known (elliptic) a posteriori error indicator,
- $\langle R_{\tilde{U}}, v \rangle$ is well-defined, non-singular, and can be estimated by a computable a posteriori indicator,
- $\langle R_{\tilde{U}}, v \rangle$ satisfy the original PDE with a modified right-hand side.

In this book, the tool to obtain such an auxiliary solution $\tilde{U}$ is the *elliptic reconstruction* technique introduced in [67] to derive a posteriori error estimates for linear semi-discrete problems. In what follows, the *elliptic reconstruction* is an operator $\mathscr{R} : V_h \mapsto V$ with $\mathscr{R}U = \tilde{U}$, which is stated in the following definition.

Definition 4.1. [67, Definition 2.1] For each $t \in (0,T]$, let $U \in V_h$ be the solution of the discrete system (4.16). The elliptic reconstruction $\tilde{U} = \mathscr{R}U \in V$ of U is defined as the solution of the stationary problem

$$a(\tilde{U}(t),v) = \langle g_h(t),v \rangle , \quad \forall v \in V, \text{ a.e. } t \in (0,T]$$

with the modified right-hand side term

$$g_h := A_h U - f_h + f,$$

where $A_h : V_h \mapsto V_h$ is the discrete version of A satisfying

$$a(U,v) = \langle A_h U, v \rangle \quad \forall v \in V_h,$$

and f_h denote the L^2-projection of f onto the space V_h.

In the sequel, we will use the *elliptic reconstruction* technique to derive a posteriori error indicators for the semi-discrete system (4.9) and the fully-discrete system (4.14) using the a posteriori error estimate (3.8) for the stationary model in Chapter 3. This approach is studied in [26] for non-stationary linear ADR equations. Here, we adopt it to the non-stationary semi-linear ADR equations.

4.3.1 A Posteriori Error Bounds for Semi-Discrete System

In order to measure the error for the semi-discrete problem, we use the $L^\infty(L^2) + L^2(H^1)$-type norm

$$\|v\|_*^2 = \|v\|_{L^\infty(0,T;L^2(\Omega))}^2 + \int_0^T |||v|||^2 dt,$$

and we also take into account the dG-norm

$$\|v\|_{dG} = \||v\|| + |v|_C,$$

where the energy norm $\||\cdot\||$ and the semi-norm $|\cdot|_C$ are defined as in (3.10) and (3.11) in Chapter 3, respectively.

We make use of the elliptic reconstruction technique in [67]. In the view of the continuous semi-discrete problem (4.5), for each $t \in (0,T]$, the elliptic reconstruction $w \in H_0^1(\Omega)$ is the unique solution of

$$a(t;w,v) + b(t;w,v) = \int_\Omega \left(f - \frac{\partial u_h}{\partial t} \right) v dx , \quad \forall v \in H_0^1(\Omega). \tag{4.17}$$

The SIPG discretization, on the other hand, of the above system reads as: for each $t \in (0,T]$, find $w_h \in C^{0,1}(0,T;V_h(\xi))$ satisfying for all $v_h \in V_h(\xi)$,

$$a_h(t;w_h,v_h) + K_h(w_h,v_h) + b_h(t;w_h,v_h) = \int_\Omega \left(f - \frac{\partial u_h}{\partial t} \right) v_h dx,$$

which implies according to (4.9) that $w_h = u_h$. Hence, the error bound to the term $\|w - u_h\|_{dG}$ can be found using the ready a posteriori error bound (3.12) for non-linear stationary problem.

Most of the steps of the construction of a posteriori error bounds for the semi-discrete system is analogous to the ones in [26]. For the error $e(t) = u(t) - u_h(t)$ of the semi-discrete problem, we set the decomposition $e(t) = \mu(t) + v(t)$ with

$$\mu(t) = u(t) - w(t) , \qquad v = w(t) - u_h(t).$$

Further, as in the stationary case, we also decompose the SIPG solution $u_h(t) \in V_h$ for each $t \in (0,T]$ as

$$u_h(t) = u_h^c(t) + u_h^r(t)$$

with $u_h^c(t) \in H_0^1(\Omega) \cap V_h$ being the conforming part of the solution and $u_h^r \in V_h$ is the remainder term. By this construction, we will have the conforming error definitions

$$e^c(t) = u(t) - u_h^c(t) , \qquad v^c(t) = w(t) - u_h^c(t). \tag{4.18}$$

Theorem 4.1. *For the time-dependent error $e = u - u_h$ of the semi-discrete system (4.9), we have*

$$\|e\|_* \lesssim \tilde{\eta}, \tag{4.19}$$

where the error estimator $\tilde{\eta}$ is defined by

$$\tilde{\eta}^2 = \|e(0)\|^2 + \int_0^T \tilde{\eta}_{S_1}^2 dt + \min \left\{ \left(\int_0^T \tilde{\eta}_{S_2}^2 dt \right)^2, \rho_T^2 \int_0^T \tilde{\eta}_{S_2}^2 dt \right\} + \max_{0 \le t \le T} \tilde{\eta}_{S_3}^2,$$

with

$$\tilde{\eta}_{S_1}^2 = \sum_{K \in \xi} \rho_K^2 \left\| f - \frac{\partial u_h}{\partial t} + \varepsilon \Delta u_h - \beta \cdot \nabla u_h - r(u_h) \right\|_{L^2(K)}^2 + \sum_{e \in \Gamma_h^0} \varepsilon^{-1/2} \rho_e \| [\varepsilon \nabla u_h] \|_{L^2(e)}^2$$

$$+ \sum_{e \in \Gamma_h} \left(\frac{\varepsilon \sigma}{h_e} + \alpha_0 h_e + \frac{h_e}{\varepsilon} \right) \| [u_h] \|_{L^2(e)}^2,$$

$$\tilde{\eta}_{S_2}^2 = \sum_{e \in \Gamma_h} h_e \left\| \left[\frac{\partial u_h}{\partial t} \right] \right\|_{L^2(e)}^2,$$

$$\tilde{\eta}_{S_3}^2 = \sum_{e \in \Gamma_h} h_e \| [u_h] \|_{L^2(e)}^2,$$

and the weight $\rho_T = min(\varepsilon^{-\frac{1}{2}}, \alpha_0^{-\frac{1}{2}})$. Note that the indicator $\tilde{\eta}_{S_1}$ is the same as the error indicator η in (3.7) for the elliptic model with a modified right-hand side and without the linear reaction term.

Proof. For any $t \in (0, T]$, let $u = u(t)$ and $u_h = u_h(t)$ be the exact solution and SIPG semi-discrete approximation of (4.4), respectively. For any $v \in H_0^1(\Omega)$, the equations (4.12) and (4.17) reads

$$\int_\Omega \frac{\partial u}{\partial t} v dx + a_h(t; u, v) + \int_\Omega r(u) v dx = \int_\Omega f v dx, \tag{4.20}$$

$$a_h(t; w, v) + \int_\Omega r(w) v dx = \int_\Omega \left(f - \frac{\partial u_h}{\partial t} \right) v dx. \tag{4.21}$$

Subtracting (4.21) from (4.20), we obtain

$$\int_\Omega \frac{\partial e}{\partial t} v dx + a_h(t; \mu, v) + \int_\Omega (r(u) - r(w)) v dx = 0. \tag{4.22}$$

Choosing $v = e^c \in H_0^1(\Omega)$ in (4.22) and using the error definitions (4.18), we get

$$\int_\Omega \frac{\partial e^c}{\partial t} e^c dx + a_h(t; e^c, e^c) + \int_\Omega (r(u) - r(w)) e^c dx = \int_\Omega \frac{\partial u_h^c}{\partial t} e^c dx + a_h(t; v^c, e^c). \tag{4.23}$$

In (4.23), using the Young's inequality, and imposing the coercivity and continuity facts (4.7) and (4.8), respectively, we arrive at

$$\frac{d}{dt} \|e^c\|^2 + \|\|e^c\|\|^2 + \int_\Omega (r(u) - r(w)) e^c dx \lesssim \|v^c\|_{dG} + \left\| \frac{\partial u_h^c}{\partial t} \right\| \|e^c\|. \tag{4.24}$$

Adding and subtracting the term $r(u_h^c)$ into the integral term related to the non-linear term in (4.24), we obtain

$$\frac{d}{dt}\|e^c\|^2 + \||e^c\||^2 + \int_\Omega (r(u) - r(u_h^c))e^c dx \lesssim \|v^c\|_{dG} + \left\|\frac{\partial u_h^c}{\partial t}\right\| \|e^c\|$$

$$+ \int_\Omega (r(w) - r(u_h^c))e^c dx. \tag{4.25}$$

Now, consider the integral terms related to the non-linear term in (4.25). Using the Cauchy-Schwarz's and Young's inequalities, and the local Lipschitz condition (4.2b), we have

$$\int_\Omega (r(w) - r(u_h^c))e^c dx \leq L(S)\|v^c\|\|e^c\|$$

$$\lesssim \|v^c\|_{dG}^2. \tag{4.26}$$

Again using the Cauchy-Schwarz's and Young's inequalities, and the conditions (4.2a) and (4.2c), we obtain

$$\int_\Omega (r(u) - r(u_h^c))e^c dx \gtrsim \||e^c\||^2. \tag{4.27}$$

Thus, combining the identities (4.26) and (4.27), the inequality (4.25) becomes

$$\frac{d}{dt}\|e^c\|^2 + \||e^c\||^2 \lesssim \|v^c\|_{dG} + \left\|\frac{\partial u_h^c}{\partial t}\right\| \|e^c\|. \tag{4.28}$$

Finally, the error bound (4.19) easily follows from (4.28) and [26, Theorem 5.4].

4.3.2 A Posteriori Error Bounds for a Fully Discrete System

We consider the solutions at discrete time instances for the fully discrete system (4.14). Let any $v_h^k \in V_h^k$, let $A^k \in V_h^k$ be the unique solution of the stationary system

$$a_h(t^k; u_h^k, v_h^k) + K_h(u_h^k, v_h^k) + b_h(t^k; u_h^k, v_h^k) = \int_\Omega A^k v_h^k dx. \tag{4.29}$$

Note that for $k \geq 1$, $A^k = I_h^k f^k - (u_h^k - I_h^k u_h^{k-1})/\Delta t_k$ with I_h^k being the L^2-projection operator onto the space V_h^k. Then, the elliptic reconstruction $w^k \in H_0^1(\Omega)$ is defined as the unique solution of the stationary problem

$$a_h(t^k; w^k, v) + b(t^k; w^k, v) = \int_\Omega A^k v dx, \qquad \forall v \in H_0^1(\Omega). \tag{4.30}$$

Next, we define the time-dependent solution $u_h(t) \in V_h^k \cup V_h^{k+1}$ as a piecewise continuous functions so that on each interval $(t^{k-1}, t^k]$, $u_h(t)$ is the linear interpolation of u_h^{k-1} and u_h^k given by

$$u_h(t) = l_{k-1}(t)u_h^{k-1} + l_k(t)u_h^k \tag{4.31}$$

with the linear Lagrange interpolation functions l_{k-1} and l_k are defined on $[t^{k-1}, t^k]$. Further, we use the decomposition of each discrete solution $u_h^k = u_h^{k,c} + u_h^{k,r}$ with $u_h^{k,c} \in H_0^1(\Omega)$ and $u_h^{k,r} \in V_h^k$, as well as the errors $e^c = u - u_h^c$ and $v^k = w^k - u_h^k$. On an arbitrary interval $(t^{k-1}, t^k]$, we have

$$u_h^c(t) = l_{k-1}(t) u_h^{k-1,c} + l_k(t) u_h^{k,c}.$$

Theorem 4.2. *Through the definition (4.31), for the in time continuous error $e = u - u_h$ of the fully-discrete system (4.14), there holds*

$$\|e\|_*^2 \lesssim \eta_S^2 + \eta_T^2, \tag{4.32}$$

where the spatial estimator η_S is given by [26]

$$\eta_S^2 = \|e(0)\|^2 + \frac{1}{3} \sum_{k=1}^n \Delta t_k (\eta_{S_{1,k-1}}^2 + \eta_{S_{1,k}}^2) + \sum_{k=1}^n \Delta t_k \eta_{S_{2,k}}^2 + \max_{0 \le k \le n} \eta_{S_{3,k}}^2$$

$$+ \min\left\{ \left(\sum_{k=1}^n \Delta t_k \eta_{S_{4,k}} \right)^2, \rho_T^2 \sum_{k=1}^n \Delta t_j \eta_{S_{4,k}}^2 \right\} \tag{4.33}$$

with

$$\eta_{S_{1,k}}^2 = \sum_{K \in \xi^{k-1} \cup \xi^k} \rho_K^2 \left\| A^k + \varepsilon \Delta u_h^k - \beta^k \cdot \nabla u_h^k - r(u_h^k) \right\|_{L^2(K)}^2$$

$$+ \sum_{e \in \Gamma_h^0} \varepsilon^{-1/2} \rho_e \| [\varepsilon \nabla u_h^k] \|_{L^2(e)}^2 + \sum_{e \in \Gamma_h} \left(\frac{\varepsilon \sigma}{h_e} + \alpha_0 h_e + \frac{h_e}{\varepsilon} \right) \| [u_h^k] \|_{L^2(e)}^2,$$

$$\eta_{S_{2,k}}^2 = \sum_{K \in \xi^{k-1} \cup \xi^k} \rho_K^2 \left\| f^k - I_h^k f^k + \frac{u_h^{k-1} - I_h^k u_h^{k-1}}{\Delta t_k} \right\|_{L^2(K)}^2,$$

$$\eta_{S_{3,k}}^2 = \sum_{e \in \Gamma_h} h_e \| [u_h^k] \|_{L^2(e)}^2,$$

$$\eta_{S_{4,k}}^2 = \sum_{e \in \Gamma_h} h_e \left\| \left[\frac{u_h^k - u_h^{k-1}}{\Delta t_k} \right] \right\|_{L^2(e)}^2,$$

and the temporal estimator η_T is given by [26]

$$\eta_T^2 = \sum_{k=1}^n \int_{t_{k-1}}^{t_k} \eta_{T_{1,k}}^2 dt$$

$$+ \min\left\{ \left(\sum_{k=1}^n \int_{t_{k-1}}^{t_k} \eta_{T_{2,k}} dt \right)^2, \rho_T^2 \sum_{k=1}^n \int_{t_{k-1}}^{t_k} \eta_{T_{2,k}}^2 dt \right\} \tag{4.34}$$

with

$$\eta_{T_{1,k}}^2 = \sum_{K \in \xi^{k-1} \cup \xi^k} \varepsilon^{-1} \| l_k(\beta^k - \beta) u_h^k + l_{k-1}(\beta^{k-1} - \beta) u_h^{k-1} \|_{L^2(K)}^2,$$

$$\eta_{T_{2,k}}^2 = \sum_{K \in \xi^{k-1} \cup \xi^k} \| f - f^k + l_{k-1}(A^k - A^{k-1}) + l_k(\nabla \cdot \beta^k - \nabla \cdot \beta) u_h^k$$

$$+ l_k(\nabla \cdot \beta^{j-1} - \nabla \cdot \beta) u_h^{k-1} \|_{L^2(K)}^2.$$

Proof. On an arbitrary interval $(t^{k-1}, t^k]$, let $e = e(t) = u(t) - u_h(t)$ represents the error of the fully-discrete system (4.14), where the approximate solution $u_h(t)$ is given in (4.31). Then, the systems (4.12) and (4.30) yields for all $v \in H_0^1(\Omega)$ the error equation

$$\int_\Omega \frac{\partial e}{\partial t} v dx + a_h(t; e, v) + \int_\Omega (r(u) - r(u_h)) v dx = \int_\Omega (f - f^k) v dx$$

$$+ \int_\Omega \left(f^k - \frac{\partial u_h}{\partial t} \right) v dx - a_h(t; u_h, v) - \int_\Omega r(u_h) v dx. \tag{4.35}$$

After choosing $v = e^c$ in (4.35) and using the conditions (4.2a)-(4.2c) for the non-linear term, as it was done in the proof of the semi-discrete case in Section 4.3.1, the error bound (4.32) easily follows from [26, Theorem 6.5].

4.3.3 Adaptive Algorithm

The space-time adaptive algorithm (see Fig. 4.1) for the non-stationary semi-linear model of type (1.1) is based on the residual-based a posteriori estimations of the elliptic problem in Chapter 3. The algorithm starts with an initial uniform mesh in space and with a given initial sufficiently large time step. At each time step, the space and time-step are adaptively updated according to the user defined tolerances **ttol** for time-step refinement, and **stol**$^+$ and **stol**$^-$ for spatial mesh, whereas the former corresponded to refinement and the latter corresponded to coarsening. We do not need a temporal tolerance corresponding to the time-step coarsening, since we start with a uniform distribution of $[0,T]$ with sufficiently large time-steps. Thus, it is sufficient just halving only the time intervals which violate the temporal tolerance **ttol**. Both the refinement and coarsening processes in space are determined by the indicator $\eta_{S_{1,k}}$ appearing in the spatial estimator (4.33), which is indeed an analogue to the elliptic indicator (3.7). Since the temporal estimator η_T (4.34) is not easy to compute, the adaptive refinement of the time-steps are driven by the modified temporal-indicator [26]

$$\tilde{\eta}_{T_k}^2 = \int_{t_{k-1}}^{t_k} \eta_{T_{1,k}}^2 dt + \min\{\rho_T, T\} \int_{t_{k-1}}^{t_k} \eta_{T_{2,k}}^2 dt,$$

sum of which gives a bound for the temporal estimator η_T^2.

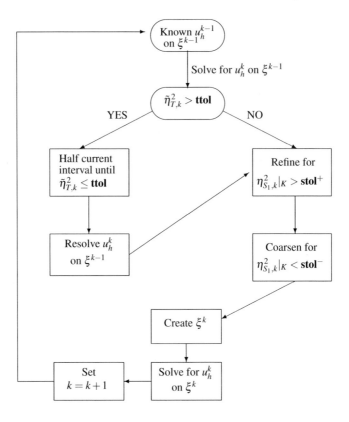

Fig. 4.1: Adaptive algorithm chart

Although the adaptive algorithm, Fig. 4.1, stands for a single equation of the system (4.4), it is easy to extend the algorithm to the coupled systems. For example when we consider a system of two equations, the temporal-indicators $\tilde{\eta}^1_{T_k}$, $\tilde{\eta}^2_{T_k}$ and the spatial indicators $\eta^1_{S_{1,k}}$, $\eta^2_{S_{1,k}}$ corresponding to the unknowns u_1 and u_2 are computed, respectively. For time-step adaptation, we require the time step condition to be satisfied for both temporal-indicators, i.e., $\tilde{\eta}^1_{T_k} \leq \textbf{ttol}$ and $\tilde{\eta}^2_{T_k} \leq \textbf{ttol}$. On the other hand, in the refinement process, we take the set of elements which is the union of the sets of the elements satisfying $\eta^1_{S_{1,k}} > \textbf{stol}^+$ and $\eta^2_{S_{1,k}} > \textbf{stol}^+$. A similar procedure applied in coarsening, but not including any elements which are selected to be refined. Numerical studies demonstrate that the adaptive algorithm is very capable of resolving the layers in space and in time.

4.4 Solution of a Fully Discrete System

In this section, we discuss the solution of the fully-discrete system (4.14) on an arbitrary k^{th} time-step for all $k = 1, 2, \ldots, n$. For consistency of notation, we consider the system (4.14) on an arbitrary k^{th} time-step without superscripts for the time-step of the form

$$\int_{\Omega} \frac{u_h - w_h}{\Delta t} v_h dx + a_h(u_h, v_h) + K_h(u_h, v_h)$$
$$+ b_h(u_h, v_h) = \int_{\Omega} f v_h dx, \quad \forall v_h \in V_h \tag{4.36}$$

where we have set $u_h := u_h^k$, $w_h := u_h^{k-1}$, $v_h := v_h^k$, $f := f^k$, $\Delta t := \Delta t_k$, $a_h(u_h, v_h) := a_h(t^k; u_h^k, v_h^k)$, $K_h(u_h, v_h) := K_h(u_h^k, v_h^k)$, $b_h(u_h, v_h) := b_h(t^k; u_h^k, v_h^k)$ and $V_h := V_h^k$. The approximate solution u_h and the known solution (from the previous time-step) w_h of (4.36) have the form

$$u_h = \sum_{i=1}^{Nel} \sum_{l=1}^{Nloc} u_l^i \phi_l^i, \quad w_h = \sum_{i=1}^{Nel} \sum_{l=1}^{Nloc} w_l^i \phi_l^i, \tag{4.37}$$

where ϕ_l^i's are the basis polynomials spanning the space V_h. $\mathbf{U} = \{u_l^i\}$ and $\mathbf{W} = \{w_l^i\}$ are the vectors of unknown and known coefficients, respectively. Using the notations and definitions in Section 3.1.1, the discrete residual of the system (4.36) in matrix vector form is given by

$$R(\mathbf{U}) = M\mathbf{U} - M\mathbf{W} + \Delta t(S\mathbf{U} + \mathbf{b}(\mathbf{U}) - \mathbf{L}) = 0, \tag{4.38}$$

where M is the mass matrix, S is the stiffness matrix corresponding to the bilinear form $a_h(u_h, v_h) + K_h(u_h, v_h)$, $\mathbf{b}(\mathbf{U})$ is the vector function of $\mathbf{U}$ related to the non-linear form $b_h(u_h, v_h)$ and $\mathbf{L}$ is the vector to the linear form $l_h(v_h)$.

Next, we solve the system (4.38) by Newton method, starting with an initial guess $\mathbf{U}^{(0)}$. As starting vector we take $\mathbf{U}^{(0)} = \mathbf{W}$, i.e. the known solution from the previous time-step and for $i = 0, 1, 2, \ldots$, we solve the system

$$J^i \delta \mathbf{U}^{(i)} = -R(\mathbf{U}^{(i)}),$$
$$\mathbf{U}^{(i+1)} = \mathbf{U}^{(i)} + \delta \mathbf{U}^{(i)} \tag{4.39}$$

until a prescribed tolerance is satisfied. In (4.39), the sparse matrix $J^i = M + \Delta t(S + J_b^i)$ denotes the Jacobian matrix of the residual function $R(\mathbf{U})$ at the current iterate $\mathbf{U}^{(i)}$, and J_b^S stands for the Jacobian matrix to the vector function $\mathbf{b}(\mathbf{U})$ at the current iterate $\mathbf{U}^{(i)}$.

4.5 Numerical Examples

In this section, we discuss some numerical studies that demonstrate the performance of the space-time adaptive algorithm. All the computations are implemented on MATLAB-R2014a. In these problems, by the term "very coarse initial mesh", we mean an initial mesh which is formed, for instance on $\Omega = (0,1)^2$, by dividing the region with $\Delta x_1 = \Delta x_2 = 0.5$ leading to 8 triangular elements and 48 DoFs for quadratic elements.

We consider as the benchmark problem, a polynomial non-linearity with a non-moving internal layer. We expect that the effectivity indices lie in a small band for different diffusion parameters meaning that our estimators are robust in the system Péclet number. For space-time adaptive algorithm we use the weighted DoFs

$$\text{Weighted DoFs} = \frac{1}{T} \sum_{k=1}^{n} \Delta t_k \lambda_k,$$

where λ_k denotes the total number of DoFs on the union mesh $\xi^{k-1} \cup \xi^k$. Since the first example has a non-moving internal layer, a monotonic increase in the DoFs is expected as time progresses. For problems with moving layers as in Example 4.5.2-4.5.3 we expect that the refinement and coarsening takes places, leading to oscillations in time vs DoFs plots. In Example 4.5.2, we also test the performance of our algorithm for a coupled system. Finally we consider in Example 4.5.4, a real problem from geosciences, reaction mechanism in porous media with internal layers at different locations due to high-permeability rocks.

4.5.1 A Problem with Polynomial Non-Linearity (Benchmark of the Algorithm)

The first example is taken from [16] with a polynomial non-linear term

$$u_t + \beta \cdot \nabla u - \varepsilon \Delta u + u^4 = f \quad \text{in } \Omega = (0,1)^2$$

with the advection field $\beta = (2,3)^T$ and the diffusion coefficient $\varepsilon = 10^{-6}$. The source function f and the Dirichlet boundary condition are chosen so that the exact solution is given by

$$u(\mathbf{x},t) = 16\sin(\pi t)x_1(1-x_1)x_2(1-x_2)$$
$$[0.5 + \pi^{-1}\arctan(2\varepsilon^{-1/2}(0.25^2 - (x_1 - 0.5)^2 - (x_2 - 0.5)^2))].$$

We start by demonstrating the decrease of errors by uniform space-time refinement using linear dG elements. In Fig. 4.2, the expected first-order convergence in space and time is shown.

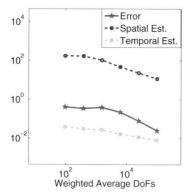

Fig. 4.2: Example 4.5.1: Decays of estimators and errors for uniform space-time

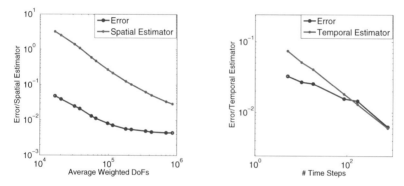

Fig. 4.3: Example 4.5.1: Error vs. spatial (left) and temporal (right) estimators for $\varepsilon = 10^{-6}$

In the following, we use quadratic dG elements. We investigate the performance of the spatial estimator by fixing the temporal time-step $\Delta t = 0.002$ so that the temporal error is dominated by the spatial error. Then, we reduce the spatial estimator tolerance **stol**$^+$ from 10^{-1} to 10^{-6}. The convergence rate and the spatial estimators are similar as illustrated in Fig. 4.3, left, for $\varepsilon = 10^{-6}$. Fig. 4.4 shows the spatial effectivity indices and the decrease of the spatial estimators for various diffusion constants ε. One can see that the effectivity indices converges asymptotically to a small band, as the results in [26] for linear problems, showing the robustness of the spatial estimator.

To investigate the performance of the temporal estimator, we take a sufficiently fine spatial mesh so that the spatial error is dominated by the temporal error. Then we reduce the temporal estimator tolerance **ttol** in the range $[10^{-1}, 10^{-6}]$. In Fig. 4.5, the temporal effectivity indices and the decrease of the temporal estimators are not

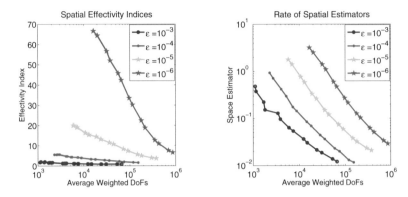

Fig. 4.4: Example 4.5.1: Spatial effectivity indices (left) and estimators (right)

affected by ε, and effectivity indices are almost the same within the band 1-2, showing the robustness of the temporal estimator.

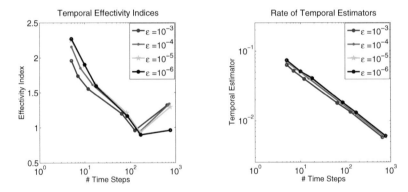

Fig. 4.5: Example 4.5.1: Temporal effectivity indices (left) and estimators (right)

Finally, we apply the space-time adaptive algorithm with the tolerances **ttol** $= 10^{-3}$, **stol**$^{+} = 3 \times 10^{-4}$ and **stol**$^{-} = 3 \times 10^{-7}$. We prepare a very coarse initial spatial mesh and a uniform partition of the time interval $[0, 0.5]$ with the step-size $\Delta t = 0.25$ until the user defined tolerances **ttol** and **stol**$^{+}$ are satisfied. The adaptive mesh at the final time $T = 0.5$ is shown in Fig. 4.6. In Fig. 4.8 on the right, the change of the time-steps is shown, whereas the change in the DoFs is illustrated in Fig. 4.8 on the left. Since the layers in the problem do not move as the time progresses, the number of DoFs increases monotonically by the spatial grid refinement. In Fig. 4.7, it is shown that all the oscillations are damped out by adaptive algorithm using less DoFs compared with the uniform refinement.

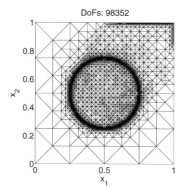

Fig. 4.6: Example 4.5.1: Adaptive mesh

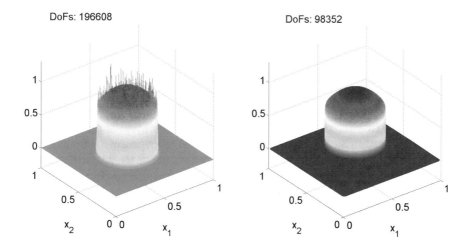

Fig. 4.7: Example 4.5.1: Uniform (left) and adaptive (right) solutions at T=0.5

4.5.2 Coupled Problem with Polynomial Non-Linearity

The next example is a coupled non-linear problem taken from [17].

$$\frac{\partial u_i}{\partial t} - \varepsilon \Delta u_i + \beta_i \cdot \nabla u_i + u_1 u_2 = f_i, \quad i = 1,2$$

on $\Omega = (0,1)^2$ with the advection fields $\beta_1 = (1,0)^T$ and $\beta_2 = (-1,0)^T$, and the diffusion constant $\varepsilon = 10^{-5}$. The Dirichlet boundary conditions, initial conditions and the load functions f_i are chosen so that the exact solutions are

$$u_1(\mathbf{x},t) = \frac{1}{2} \left(1 - \tanh \frac{2x_1 - 0.2t - 0.8}{\sqrt{5\varepsilon}} \right),$$

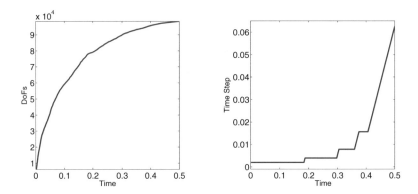

Fig. 4.8: Example 4.5.1: Evolution of DoFs (left) and time-steps Δt (right)

$$u_2(\mathbf{x},t) = \frac{1}{2}\left(1 + \tanh\frac{2x_1 + 0.2t - 0.9}{\sqrt{5\varepsilon}}\right).$$

Again we use quadratic dG elements. We prepare an initial mesh, Fig. 4.10 on the left, starting with a very coarse spatial mesh and a uniform partition of the time interval $[0,1]$ with the step-size $\Delta t = 0.1$ until the user defined tolerances **ttol** $= 10^{-3}$ and **stol**$^+ = 10^{-1}$ are satisfied. Here, two sharp fronts move towards each other and then mix directly after the time $t = 0.1$, Fig. 4.9. The movement of the fronts are also visible in Fig. 4.10 claiming that refinement/coarsening of the adaptive algorithm works well. We see that the sharp fronts in the cross-wind direction $x_2 = 0.5x_1 + 0.75$ are almost damped out. Moreover, Fig. 4.10-4.11 show that both the spatial and temporal estimators catch the time where the two sharp fronts mix.

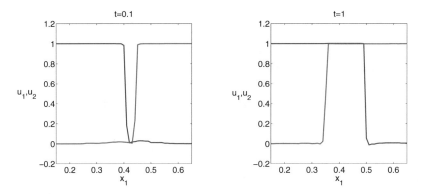

Fig. 4.9: Example 4.5.2: Cross-section plots in the cross-wind direction at $t = 0.1$ (left) and $t = 1$ (right)

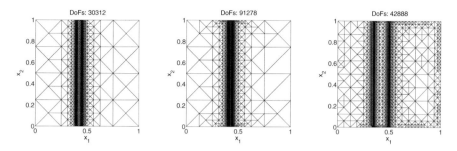

Fig. 4.10: Example 4.5.2: Adaptive meshes at $t = 0$, $t = 0.1$ and $t = 1$ (from left to right)

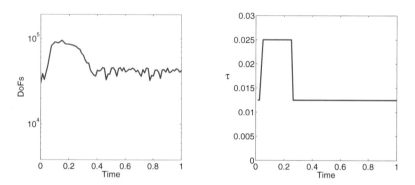

Fig. 4.11: Example 4.5.2: Evolution of DoFs (left) and time-steps Δt (right)

4.5.3 Non-linear ADR equation in Homogeneous Porous Media

We consider the advection-diffusion-reaction (ADR) equation in [85] with polynomial type non-linear reaction

$$\frac{\partial u}{\partial t} - \varepsilon \Delta u + \beta \cdot \nabla u + \gamma u^2 (u - 1) = 0 \quad \text{in } \Omega \times (0, T]$$

with $\Omega = (0, 1)^2$. We take as in [85] the homogeneous dispersion tensor as $\varepsilon = 10^{-4}$, the velocity field $\beta = (-0.01, -0.01)^T$ and $\gamma = 100$. The initial and boundary conditions are chosen by the exact solution

$$u(\mathbf{x}, t) = [1 + \exp(a(x_1 + x_2 - bt) + a(b - 1))]^{-1},$$

where $a = \sqrt{\gamma/(4\varepsilon)}$ and $b = -0.02 + \sqrt{\gamma\varepsilon}$. The problem is a transport of a front in homogeneous porous media. We simulate the given problem for the final time $T = 1$, and with quadratic dG elements. We begin a very coarse spatial initial mesh and a uniform partition of the time interval $[0,1]$ with the step-size $\Delta t = 0.25$ until the user defined tolerances **ttol** $= 3 \times 10^{-3}$ and **stol**$^{+} = 10^{-3}$ are satisfied. In Fig. 4.12, the adaptive meshes and solution profiles are shown at the times $t = \{0.2, 0.6, 1\}$, where the movement of the front can be seen. The time vs DoFs and time vs time step-size plots in Fig. 4.13 indicate clearly the oscillations in DoFs and time-steps due to the movement of the front.

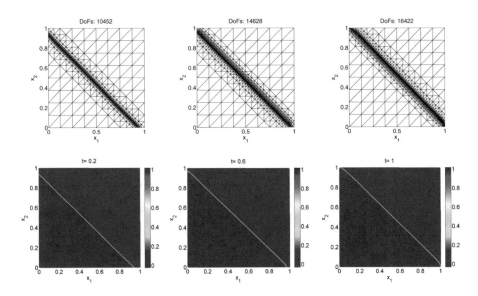

Fig. 4.12: Example 4.5.3: Adaptive meshes (top) and solution profiles (bottom) at $t = 0.2$, $t = 0.6$ and $t = 1$ (from left to right)

4.5.4 Non-linear ADR equation in Deterministic Heterogeneous Porous Media

As a last example we consider the ADR equation in [85] with Monod or Langmuir isotherm type non-linear reaction terms

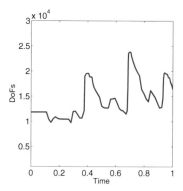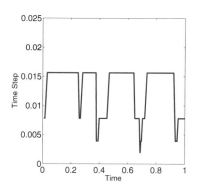

Fig. 4.13: Example 4.5.3: Evolution of DoFs (left) and time-steps Δt (right)

$$
\begin{aligned}
\frac{\partial u}{\partial t} - \nabla \cdot (\varepsilon \nabla u) + \beta(x) \cdot \nabla u + \frac{u}{1+u} &= 0 && \text{in } \Omega \times (0,T], \\
u(x,t) &= 1 && \text{on } \Gamma^D \times [0,T], \\
-\varepsilon \nabla u(x,t) \cdot \mathbf{n} &= 0 && \text{on } (\partial\Omega \setminus \Gamma^D) \times [0,T], \\
u(x,0) &= 0 && \text{in } \Omega
\end{aligned}
$$

with $\Omega = (0,3) \times (0,2)$ and $\Gamma^D = \{0\} \times [0,2]$. The problem represents reaction in porous media, the transport in a highly idealized fracture pattern. Here ε stands for the heterogeneous dispersion tensor given by

$$
\varepsilon = \begin{bmatrix} 10^{-3} & 0 \\ 0 & 10^{-4} \end{bmatrix}.
$$

The velocity field $\beta(x)$ is determined via Darcy's law

$$
\beta = -\frac{k(x)}{\mu} \nabla p,
$$

where p is the fluid pressure, μ is the fluid viscosity and $k(x)$ is the permeability of the porous medium. Using the mass conservation property $\nabla \cdot \beta(x) = 0$ under the assumption that rock and fluids are incompressible, the velocity field $\beta(x)$ is computed by solving the system

$$
\begin{aligned}
\nabla \cdot \left(\frac{k(x)}{\mu} \nabla p \right) &= 0 && \text{in } \Omega, \\
p &= 1 && \text{on } \{0\} \times [0,2], \\
p &= 0 && \text{on } \{3\} \times [0,2], \\
-k(x)\nabla p \cdot \mathbf{n} &= 0 && \text{on } (0,3) \times \{0,2\}.
\end{aligned}
$$

We simulate the given problem for the final time $T = 1$ using linear dG elements. We take the fluid viscosity $\mu = 0.1$, and the permeability field as in [85] with three parallel streaks of the permeability of which are 100 times greater than the permeability of the surrounding domain. In Fig. 4.14 on the left, the flow is canalized from the lower-permeability rocks into the high-permeability ones, in Fig. 4.14 on the right. For the adaptive procedure, we again take a coarse spatial initial mesh and a uniform partition of the time interval $[0, 1]$ with the step-size $\Delta t = 0.05$ until the user defined tolerances **ttol** $= 10^{-3}$ and **stol**$^+ = 3 \times 10^{-4}$ are satisfied. Fig. 4.15-4.16 show the adaptive meshes and concentrations at $t = 0.3$ and $t = 1$, where we can clearly see the flow-focusing due to the high-permeability.

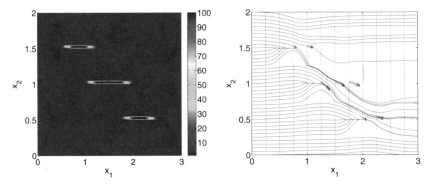

Fig. 4.14: Example 4.5.4: Permeability field (left) and velocity streamlines (right)

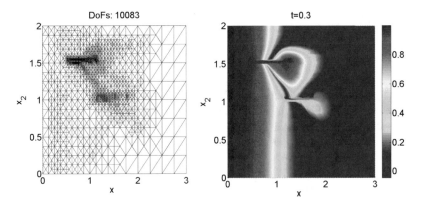

Fig. 4.15: Example 4.5.4: Adaptive mesh (left) and concentration (right) at $t = 0.3$

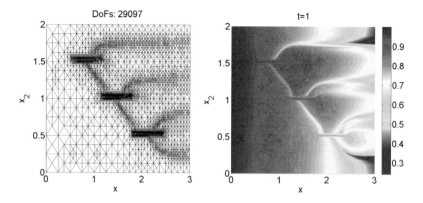

Fig. 4.16: Example 4.5.4: Adaptive mesh (left) and concentration (right) at $t = 1$

In Fig. 4.17 plots for time vs DoFs and time vs time step-size are shown. We see that initially time steps are small reaching a steady time step in Fig. 4.17 on the right. The number of DoFs increases (refinement dominates coarsening) monotonically after the meet of first high-permeability rock until the meet of third high-permeability rock and then the increase stops, Fig. 4.17 on the left. This is meaningful since there is no sharp flow canalization after the third high-permeability rock.

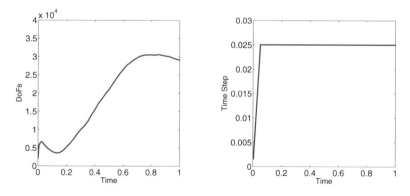

Fig. 4.17: Example 4.5.4: Evolution of DoFs (left) and time-steps Δt (right)

Chapter 5
Conclusions and Outline

In this book adaptive algorithms are developed for efficient discretization of advection dominated stationary and non-stationary semi-linear ADR equations. In order to handle the unphysical oscillations due to the advection, we have applied a symmetric interior penalty Galerkin (SIPG) method as an alternative to the well-known stabilized continuous FEM methods such as the streamlined upwind Petrov-Galerkin (SUPG) method. We have given a detailed construction of SIPG formulation on the general Poisson equation, and we have discussed the effect of the penalty parameter in Chapter 2. In Chapter 3, we gave existence and uniqueness results for stationary semi-linear ADR equations. We have shown that the space-time adaptive algorithm is robust and can resolve not only the layers produced by advection but also the sharp fronts due to the non-linear reaction as an alternate to the shock/discontinuity capturing techniques in the literature. We have also shown that adaptive dG approximations for stationary problems are more accurate than the Galerkin least squares FEMs and shock/discontinuity capturing techniques. Moreover, we have introduced an efficient iterative method, matrix reordering technique, as a preconditioner to solve the linear systems arising from the Newton's method applied to the discrete system of stationary semi-linear ADR equations.

In Chapter 4, we have considered the non-stationary semi-linear ADR, discretized in space by SIPG and in time by the backward Euler method. We proved a posteriori error bounds for the semi-discrete and fully discrete systems using $L^\infty(L^2)+L^2(H^1)$-type norm. To construct the energy-norm a posteriori error bounds for the non-stationary model, we have utilized the elliptic reconstruction technique to use a posteriori error bounds driven for stationary models in contrast to standard energy techniques which have to adapt the estimates case by case in order to compare the exact solution with numerical solution directly and are sub-optimal order in $L^\infty(L^2)$-type norms. Using driven a posteriori error estimates, we have introduced an adaptive algorithm both in space and time, in space using both refinement and coarsening. Through the numerical studies, we have demonstrated that our adaptive algorithm captures the interior and boundary layers very sharply without any significant oscillation, and also the temporal layers. In addition, we have shown the efficiency and in diffusion parameter robustness of the algorithm numerically

© Springer International Publishing Switzerland 2016
M. Uzunca, *Adaptive Discontinuous Galerkin Methods for Non-linear Reactive Flows*,
Lecture Notes in Geosystems Mathematics and Computing,
DOI 10.1007/978-3-319-30130-3_5

by demonstrating the spatial and temporal effectivity indices, and the rates of the errors and estimators. We have noted that our results are similar to those ones for non-stationary linear ADR equations in [26].

In contrast to the residual-based energy techniques, there are more popular approaches such as goal oriented and hierarchical error estimates for non-stationary problems. We will apply these adaptive techniques for non-stationary semi-linear equations in a future work. Further, in real life, non-stationary models are parameter dependent requiring many query computations, which can be solved using model order reduction techniques. As a future work, we will study reduced order modeling based on proper orthogonal decomposition (POD) to solve the non-stationary semi-linear ADR equations using adaptive dG methods, which is an open problem in the literature.

Appendix A
Matlab Tutorial

We discuss a collection of MATLAB routines using dG methods for solving and simulating steady-state ADR equations in 2D. The code employs the sparse matrix facilities of MATLAB with the coding style "vectorization" which replaces **for** loops by matrix operations. Moreover, we utilize multiple matrix multiplications *"MULTIPROD"* [66] to decrease the number of **for** loops in an efficient way.

Linear Model Problem

The general (linear) model problem used in the code is

$$\alpha u - \varepsilon \Delta u + \beta \cdot \nabla u = f \qquad \text{in } \Omega, \qquad \text{(A.1a)}$$

$$u = g^D \qquad \text{on } \Gamma^D, \qquad \text{(A.1b)}$$

$$\varepsilon \nabla u \cdot \mathbf{n} = g^N \qquad \text{on } \Gamma^N. \qquad \text{(A.1c)}$$

The domain Ω is bounded, open, convex in $\mathbb{R}^2$ with boundary $\partial \Omega = \Gamma^D \cup \Gamma^N$, $\Gamma^D \cap \Gamma^N = \emptyset$, $0 < \varepsilon \ll 1$ is the diffusivity constant, $f \in L^2(\Omega)$ is the source function, $\beta \in \left(W^{1,\infty}(\Omega) \right)^2$ is the velocity field, $g^D \in H^{3/2}(\Gamma^D)$ is the Dirichlet boundary condition, $g^N \in H^{1/2}(\Gamma^N)$ is the Neumann boundary condition and $\mathbf{n}$ denotes the unit outward normal vector to the boundary.

Using the notations in Chapter 3, the dG discretized system to the problem (A.1) combining with the upwind discretization for the convection part reads as: find $u_h \in V_h$ such that

$$a_h(u_h, v_h) = l_h(v_h), \qquad \forall v_h \in V_h, \qquad \text{(A.2)}$$

© Springer International Publishing Switzerland 2016
M. Uzunca, *Adaptive Discontinuous Galerkin Methods for Non-linear Reactive Flows*,
Lecture Notes in Geosystems Mathematics and Computing,
DOI 10.1007/978-3-319-30130-3

$$
a_h(u_h, v_h) = \sum_{K \in \xi_h} \int_K \varepsilon \nabla u_h \cdot \nabla v_h dx + \sum_{K \in \xi_h} \int_K \alpha u_h v_h dx + \sum_{K \in \xi_h} \int_K \beta \cdot \nabla u_h v_h dx
$$

$$
- \sum_{e \in \Gamma_h^0 \cup \Gamma_h^D} \int_e \{\varepsilon \nabla u_h\} \cdot [v_h] ds + \kappa \sum_{e \in \Gamma_h^0 \cup \Gamma_h^D} \int_e \{\varepsilon \nabla v_h\} \cdot [u_h] ds
$$

$$
+ \sum_{K \in \xi_h} \int_{\partial K^- \setminus \partial \Omega} \beta \cdot \mathbf{n}(u_h^{out} - u_h^{in}) v_h ds - \sum_{K \in \xi_h} \int_{\partial K^- \cap \Gamma_h^-} \beta \cdot \mathbf{n} u_h^{in} v_h ds
$$

$$
+ \sum_{e \in \Gamma_h^0 \cup \Gamma_h^D} \frac{\sigma \varepsilon}{h_e} \int_e [u_h] \cdot [v_h] ds,
$$

$$
l_h(v_h) = \sum_{K \in \xi_h} \int_K f v_h dx + \sum_{e \in \Gamma_h^D} \int_e g^D \left(\frac{\sigma \varepsilon}{h_e} v_h + \kappa \varepsilon \nabla v_h \cdot \mathbf{n} \right) ds
$$

$$
- \sum_{K \in \xi_h} \int_{\partial K^- \cap \Gamma_h^-} \beta \cdot \mathbf{n} g^D v_h ds + \sum_{e \in \Gamma_h^N} \int_e g^N v_h ds.
$$

The parameter $\sigma \in \mathbb{R}_0^+$ should be sufficiently large independent of the mesh size h and the diffusion coefficient ε. In our code, we choose the penalty parameter σ on interior edges depending on the polynomial degree k as $\sigma = 3k(k+1)$ for the SIPG and IIPG methods, whereas, we take $\sigma = 1$ for the NIPG method. On boundary edges, we take the penalty parameter as twice the penalty parameter on interior edges.

Description of MATLAB Codes

The given codes are mostly self-explanatory with comments to explain what each section of the code does. In this section, we give a line-by-line descriptions of our main code. The use of the code consists of three main parts:

1. Mesh generation,
2. Entry of user defined quantities (boundary conditions, order of basis etc.),
3. Forming and solving linear systems,
4. Plotting the solutions.

Except the last one, all the parts above, in the case of our code, take place in the m-file *Main_Linear.m* which is the main code to be used by users for linear problems without need to modify any other m-file. The last part, plotting the solutions, takes place in the m-file *dg_error.m*.

Mesh Generation

In this section, we define the data structure of a triangular mesh on a polygonal domain in $\mathbb{R}^2$. The data structure presented here is based on simple arrays [29] which are stored in a MATLAB "struct" that collects two or more data fields in one object that can then be passed to routines. To obtain an initial mesh, firstly, we define the nodes, elements, Dirichlet and Neumann conditions in the m-file *Main_Linear.m* through the lines 14–20, and we call the *getmesh* function to form the initial mesh structure *mesh*, line 22.

```
% Nodes
Nodes = [0,0;0.5,0;1,0;0,0.5;0.5,0.5;1,0.5;0,1;0.5,1;1,1];
% Elements
Elements = [4,1,5;1,2,5;5,2,6; 2,3,6;7,4,8;4,5,8;8,5,9;5,6,9];
% Dirichlet bdry edges
Dirichlet = [1,2;2,3;1,4;3,6;4,7;6,9;7,8;8,9];
% Neumann bdry edges
Neumann = [];
% Initial mesh struct
mesh = getmesh(Nodes,Elements,Dirichlet,Neumann);
```

As it can be understood, each row in the **Nodes** array corresponds to a mesh node in which the first column designates the x-coordinate of the node and the second designates the y-coordinate, and the $i-th$ row of the **Nodes** array is called the node having index i. In the **Elements** array, each row with 3 columns corresponds to a triangular element in the mesh containing the indices of the nodes forming the 3 vertices of the triangles in the counter-clockwise orientation. Finally, in the **Dirichlet** and **Neumann** arrays, each row with 2 columns corresponds to a Dirichlet and Neumann boundary edge containing the indices of the starting and ending nodes, respectively (see Fig.A.1).

The mesh "struct" in the code has the following fields:

- Nodes, Elements, Edges, intEdges, DbdEdges, NbdEdges, intEdges
- vertices1, vertices2, vertices3,
- Dirichlet, Neumann, EdgeEls, ElementsE.

which can be reached by *mesh.Nodes*, *mesh.Elements* and so on, and they are used by the other functions to form the DG construction. In line 24–26, the initial mesh is uniformly refined several times in a "for loop" by calling the function *uniformrefine*.

```
for jj=1:2
    mesh = uniformrefine(mesh);  % Refine mesh
end
```

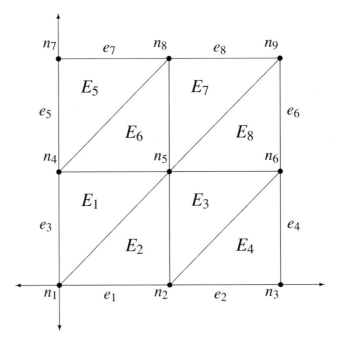

Fig. A.1: Initial mesh on the unit square $\Omega = [0,1]^2$ with nodes n_i, triangles E_j and edges e_k

User Defined Quantities

There are certain input values that have to be supplied by the user. Here, we will describe how one can define these quantities in the main code.

In lines 29–31, one determines the type of the dG method (SIPG, NIPG or IIPG) and the order of the polynomial basis to be used by the variables *method* and *degree*, respectively. According to these choices, the values of the penalty parameter and the parameter $\kappa \in \{-1,1,0\}$ defining DG method in (A.2) are set by calling the sub-function *set_parameter* in line 33.

```
% method : NIPG=1, SIPG=2, IIPG=3
method = 2;
% Degree of polynomials
degree = 1;
% Set up the problem
[penalty,kappa] = set.parameter(method,degree);
```

The next step is to supply the problem parameters. In line 89–102, the diffusion constant ε, the advection vector β and the linear reaction term α are defined via the sub-functions *fdiff*, *fadv* and *freact*, respectively.

```
function diff = fdiff(x,y)
    diff = (1e-6).*ones(size(x));
end

% Advection
function [adv1,adv2] = fadv(x,y)
    adv1 = (1/sqrt(5))*ones(size(x));
    adv2 = (2/sqrt(5))*ones(size(x));
end

% Linear reaction
function react = freact(x,y)
    react = ones(size(x));
end
```

The exact solution (if it exists) and the source function f are defined in lines 115–162 via the sub-functions *fexact* and *fsource*, respectively. Finally, in lines 168–179, the boundary conditions are supplied via the sub-functions *DBCexact* and *NBCexact*.

```
% Drichlet Boundary Condition
function DBC=DBCexact(fdiff,x,y)
    % Evaluate the diffusion function
    diff = feval(fdiff,x,y);
    % Drichlet Boundary Condition
    DBC = 0.5*(1-tanh((2*x-y-0.25)./(sqrt(5*diff))));
end

% Neumann Boundary Condition
function NC = NBCexact(mesh,fdiff,x,y)
    %Neumann Boundary Condition
    NC = zeros(size(x));
end
```

Forming and Solving Linear Systems

To form the linear systems, firstly, let us rewrite the discrete dG scheme (A.2) as

$$a_h(u_h, v_h) := D_h(u_h, v_h) + C_h(u_h, v_h) + R_h(u_h, v_h) = l_h(v_h), \qquad \forall v_h \in V_h, \quad \text{(A.3)}$$

where the forms $D_h(u_h, v_h)$, $C_h(u_h, v_h)$ and $R_h(u_h, v_h)$ corresponding to the diffusion, convection and linear reaction parts of the problem, respectively

$$D_h(u_h, v_h) = \sum_{K \in \xi_h} \int_K \varepsilon \nabla u_h \cdot \nabla v_h dx + \sum_{e \in \Gamma_0 \cup \Gamma_D} \frac{\sigma \varepsilon}{h_e} \int_e [u_h] \cdot [v_h] ds$$

$$- \sum_{e \in \Gamma_0 \cup \Gamma_D} \int_e \{\varepsilon \nabla u_h\} \cdot [v_h] ds + \kappa \sum_{e \in \Gamma_0 \cup \Gamma_D} \int_e \{\varepsilon \nabla v_h\} \cdot [u_h] ds$$

$$C_h(u_h, v_h) = \sum_{K \in \xi_h} \int_K \beta \cdot \nabla u_h v_h dx$$

$$+ \sum_{K \in \xi_h} \int_{\partial K \setminus \partial \Omega} \beta \cdot \mathbf{n}(u_h^{out} - u_h^{in}) v_h ds - \sum_{K \in \xi_h} \int_{\partial K^- \cap \Gamma^-} \beta \cdot \mathbf{n} u_h^{in} v_h ds$$

$$R_h(u_h, v_h) = \sum_{K \in \xi_h} \int_K \alpha u_h v_h dx$$

$$l_h(v_h) = \sum_{K \in \xi_h} \int_K f v_h dx + \sum_{e \in \Gamma_D} \int_e g^D \left(\frac{\sigma \varepsilon}{h_e} v_h + \kappa \varepsilon \nabla v_h \cdot \mathbf{n} \right) ds$$

$$- \sum_{K \in \xi_h} \int_{\partial K^- \cap \Gamma^-} \beta \cdot \mathbf{n} g^D v_h ds + \sum_{e \in \Gamma_N} \int_e g^N v_h ds.$$

For a set of basis functions $\{\phi_i\}_{i=1}^N$ spanning the space V_h, the discrete solution $u_h \in V_h$ is of the form

$$u_h = \sum_{j=1}^N v_j \phi_j \tag{A.5}$$

where $v = (v_1, v_2, \ldots, v_N)^T$ is the unknown coefficient vector. After substituting (A.5) into (A.3) and taking $v_h = \phi_i$, we get for $i = 1, 2, \ldots, N$, the linear systems of equations

$$\sum_{j=1}^N v_j D_h(\phi_j, \phi_i) + \sum_{j=1}^N v_j C_h(\phi_j, \phi_i) + \sum_{j=1}^N v_j R_h(\phi_j, \phi_i) = l_h(\phi_i). \tag{A.6}$$

Thus, for $i = 1, 2, \ldots, N$, to form the linear system in matrix-vector form, we need the matrices $D, C, R \in \mathbb{R}^{N \times N}$ related to the terms including the forms D_h, C_h and R_h in (A.6), respectively, satisfying

$$D v + C v + R v = F$$

with the unknown coefficient vector v and the vector $F \in \mathbb{R}^N$ related to the linear rhs functionals $l_h(\phi_i)$ such that $F_i = l_h(\phi_i)$, $i = 1, 2, \ldots, N$. In the code *Main_Linear.m*, all the matrices D, C, R and the vector F are obtained by calling the function *global_system* in lines 36–37, in which the sub-functions introduced in the previous subsection are used. In line 39, we set the stiffness matrix, *Stiff*, as the sum of the obtained matrices and we solve the linear system for the unknown coefficient vector *coef*:= v.

```
% Compute global matrices and rhs global vector
[D,C,R,F] = global.system(mesh,@fdiff,@fadv,@freact,[],...
            @fsource,@DBCexact,@NBCexact,penalty,kappa,degree);

Stiff = D + C + R;    % Stiffness matrix

coef = Stiff\F;   % Solve the linear system
```

Plotting Solutions

After solving the problem for the unknown coefficient vector, the solutions are plotted via the function *dg_error* in line 75, and also the L^2-error between the exact and numerical solution is computed.

```
% Compute L2-error and plot the solution
[l2err,hmax] = dg.error(coef,mesh,@fexact,@fdiff,degree);
```

Models with Non-Linear Reaction Mechanisms

The model problem in this case is

$$\alpha u - \varepsilon \Delta u + \beta \cdot \nabla u + r(u) = f \qquad \text{in } \Omega, \qquad (A.7a)$$
$$u = g^D \qquad \text{on } \Gamma^D, \qquad (A.7b)$$
$$\varepsilon \nabla u \cdot \mathbf{n} = g^N \qquad \text{on } \Gamma^N. \qquad (A.7c)$$

which arises from the time discretization of the time-dependent non-linear diffusion-convection-reaction equations. Here, the coefficient of the linear reaction term, $\alpha > 0$, stand for the temporal discretization, corresponding to $1/\Delta t$, where Δt is the discrete time-step. The model (A.7) differs from the model (A.1) by the additional non-linear term $r(u)$.

To solve the non-linear problems, we use the m-file *Main_Nonlinear* which is similar to the m-file *Main_Linear*, but now we use Newton iteration to solve for $i = 1, 2, \ldots, N$ the non-linear system of equations

$$\sum_{j=1}^{N} v_j D_h(\phi_j, \phi_i) + \sum_{j=1}^{N} v_j C_h(\phi_j, \phi_i) + \sum_{j=1}^{N} v_j R_h(\phi_j, \phi_i) + \int_{\Omega} r(u_h)\phi_i dx = l_h(\phi_i).$$
$$(A.8)$$

Similar to the linear case, the above system leads to the matrix-vector form

$$Dv + Cv + Rv + H(v) = F,$$

where, in addition to the matrices $D, C, R \in \mathbb{R}^{N \times N}$ and the vector $F \in \mathbb{R}^N$, we also need the vector $H \in \mathbb{R}^N$ related to the non-linear term such that

$$H_i(v) = \int_\Omega r \left(\sum_{j=1}^N v_j \phi_j \right) \phi_i dx, \quad i = 1, 2, \ldots, N.$$

For an initial guess $v^0 = (v_1^0, v_2^0, \ldots, v_N^0)^T$, we solve the system

$$
\begin{aligned}
J^k w^k &= -R^k \\
v^{k+1} &= w^k + v^k, \quad k = 0, 1, 2, \ldots
\end{aligned}
\tag{A.9}
$$

until a user defined tolerance is satisfied. In (A.9), R^k and J^k denote the vector of system residual and its Jacobian matrix at the current iterate v^k, respectively, given by

$$
\begin{aligned}
R^k &= Sv^k + H(v^k) - F \\
J^k &= S + J_H(v^k)
\end{aligned}
$$

where $J_H(v^k)$ is the Jacobian matrix of the non-linear vector H at v^k

$$
J_H(v^k) = \begin{bmatrix}
\dfrac{\partial H_1(v^k)}{\partial v_1^k} & \dfrac{\partial H_1(v^k)}{\partial v_2^k} & \cdots & \dfrac{\partial H_1(v^k)}{\partial v_N^k} \\
\vdots & \ddots & & \vdots \\
\dfrac{\partial H_N(v^k)}{\partial v_1^k} & \dfrac{\partial H_N(v^k)}{\partial v_2^k} & \cdots & \dfrac{\partial H_N(v^k)}{\partial v_N^k}
\end{bmatrix}.
$$

In the code *Main_Nonlinear*, obtaining the matrices D, C, R and the rhs vector F is similar to the linear case, but now we give the function handle *freact_nonlinear*, which is a sub-function in the main file *Main_Nonlinear*, lines 105–110, as an input to compute the source function.

```
% Compute global matrices and rhs global vector
[D,C,R,F] = global.system(mesh,@fdiff,@fadv,@freact,...
                @freact.nonlinear,@fsource,@DBCexact,@NBCexact,...
                penalty,kappa,degree);
```

```
% Non-linear reaction
function [r,dr] = freact.nonlinear(u)
  % Value of the non-linear reaction term
  r = u.^2;
  % Value of the Jacobian of the non-linear reaction term
  dr = 2*u;
end
```

In line 43, we initialize the initial guess for Newton iteration, and we solve the nonlinear system in lines 47–72. To obtain the non-linear vector H and its Jacobian J_H at the current iterate, we call the function *nonlinear_global* in line 52. It uses the function handle *freact_nonlinear* and has to be supplied by user for $r(u)$ and $r'(u)$.

```
% Newton iteration
noi=0;
for ii=1:50
    noi=noi+1;

    % Compute the non linear vector and its Jacobian matrix at
    % the current iterate
    [H,JH] = nonlinear.global(coef,mesh,@freact.nonlinear,degree);

    % Form the residual of the system
    Res = Stiff*coef + H - F;

    % Form the Jacobian matrix of the system
    % (w.r.t. unknown coefficients coef)
    J = Stiff + JH ;

    % Solve the linear system for the correction "w"
    w = J \ (-Res);

    % Update the iterate
    coef = coef + w;

    % Check the accuracy
    if norm(J*w+Res) < 1e-20
        break;
    end

end
```

Full Version of the Code *Main_Nonlinear*

```
1   % This routine solves the diffusion-convection-reaction equation
2   %
3   %     \alpha u - \epsilon*\Delta u + b\dot\nabla u + r(u) = f
4   %
5   % using DG-FEM.
6
7   function Main_Nonlinear()
8   clear all
9   clc
10
11  % Generate the mesh
12
13  % Nodes
14  Nodes = [0,0;0.5,0;1,0;0,0.5;0.5,0.5;1,0.5;0,1;0.5,1;1,1];
15  % Elements
16  Elements = [4,1,5;1,2,5;5,2,6; 2,3,6;7,4,8;4,5,8;8,5,9;5,6,9];
17  % Dirichlet bdry edges
18  Dirichlet = [1,2;2,3;1,4;3,6;4,7;6,9;7,8;8,9];
19  % Neumann bdry edges
20  Neumann = [];
21  % Initial mesh struct
22  mesh = getmesh(Nodes,Elements,Dirichlet,Neumann);
23
24  for jj=1:2
25      mesh = uniformrefine(mesh);   % Refine mesh
26  end
27
28  % method : NIPG=1, SIPG=2, IIPG=3
29  method = 2;
30  % Degree of polynomials
31  degree = 1;
32  % Set up the problem
33  [penalty,kappa] = set_parameter(method,degree);
34
35  % Compute global matrices and rhs global vector
36  [D,C,R,F] = global_system(mesh,@fdiff,@fadv,@freact,...
37                    @freact_nonlinear,@fsource,@DBCexact,@NBCexact,...
38                    penalty,kappa,degree);
39
40  Stiff = D + C + R;   % Stiffness matrix
41
42  % Initial guess for Newton iteration
43  coef = zeros(size(Stiff,1),1);
44
45  % Newton iteration
46  noi=0;
47  for ii=1:50
48      noi=noi+1;
49
```

```
50      % Compute the non-linear vector and its Jacobian matrix at
51      % the current iterate
52      [H,JH] = nonlinear.global(coef,mesh,@freact.nonlinear,degree);
53
54      % Form the residual of the system
55      Res = Stiff*coef + H - F;
56
57      % Form the Jacobian matrix of the system
58      % (w.r.t. unknown coefficients coef)
59      J = Stiff + JH ;
60
61      % Solve the linear system for the correction "w"
62      w = J \ (-Res);
63
64      % Update the iterate
65      coef = coef + w;
66
67      % Check the accuracy
68      if norm(J*w+Res) < 1e-20
69          break;
70      end
71
72  end
73
74  % Compute L2 error and plot the solution
75  [l2err,hmax] = dg.error(coef,mesh,@fexact,@fdiff,degree);
76
77  % Degree of freedom
78  dof = size(mesh.Elements,1)*(degree+1)*(degree+2)*0.5;
79
80  fprintf('  DoFs       h.max            L2-error      #it\n')
81
82  fprintf('%6d      %5.3f       %5.3e     %d\n',dof,hmax,l2err,noi);
83
84  end
85
86  %% Define diffusion, advection, and reaction as subfunctions
87
88  % Diffusion
89  function diff = fdiff(x,y)
90      diff = (1e-6).*ones(size(x));
91  end
92
93  % Advection
94  function [adv1,adv2] = fadv(x,y)
95      adv1 = (1/sqrt(5))*ones(size(x));
96      adv2 = (2/sqrt(5))*ones(size(x));
97  end
98
99  % Linear reaction
100 function react = freact(x,y)
101     react = ones(size(x));
102 end
103
```

```
104   % Non-linear reaction
105   function [r,dr] = freact_nonlinear(u)
106    % Value of the non-linear reaction term
107     r = u.^2;
108    % Value of the Jacobian of the non-linear reaction term
109     dr = 2*u;
110   end
111
112   %% Define exact solution and force as subfunctions
113
114   % Exact solution
115   function [yex,yex_x,yex_y] = fexact(fdiff,x,y)
116     % Evaluate the diffusion function
117     diff  = feval(fdiff,x,y);
118     % Exact value
119     yex = 0.5*(1-tanh((2*x-y-0.25)./(sqrt(5*diff))));
120     % First derivative value wrt x
121     yex_x = (-1./(sqrt(5*diff))).*...
122         (sech((2*x-y-0.25)./(sqrt(5*diff)))).^2;
123     % First derivative value wrt y
124     yex_y=((0.5)./(sqrt(5*diff))).*...
125         (sech((2*x-y-0.25)./(sqrt(5*diff)))).^2;
126   end
127
128   % Force function
129   function source = fsource(fdiff,fadv,freact,freact_nonlinear,x,y)
130       % Evaluate the diffusion function
131       diff = feval(fdiff,x,y );
132       % Evaluate the advection function
133       [adv1,adv2] = feval(fadv,x, y );
134       % Evaluate the reaction function
135       reac = feval(freact,x,y);
136
137       % Exact value
138       yex = 0.5*(1-tanh((2*x-y-0.25)./(sqrt(5*diff))));
139       % First derivative value wrt x
140       yex_x = (-1./(sqrt(5*diff))).*...
141           (sech((2*x-y-0.25)./(sqrt(5*diff)))).^2;
142       % First derivative value wrt y
143       yex_y=((0.5)./(sqrt(5*diff))).*...
144           (sech((2*x-y-0.25)./(sqrt(5*diff)))).^2;
145       % Second derivative value wrt x
146       yex_xx = ((0.8)./diff).*tanh((2*x-y-0.25)./...
147           (sqrt(5*diff))).*(sech((2*x-y-0.25)./(sqrt(5*diff)))).^2;
148       % Second derivative value wrt y
149       yex_yy = ((0.2)./diff).*tanh((2*x-y-0.25)./...
150           (sqrt(5*diff))).*(sech((2*x-y-0.25)./(sqrt(5*diff)))).^2;
151
152       % Value of non-linear reaction (if exists)
153       if ~isempty(freact_nonlinear)
154           nonlin_reac = freact_nonlinear(yex);
155       else
156           nonlin_reac = 0;
157       end
```

```
158        % Value of the source function
159        source = -diff.*(yex_xx+yex_yy)+(adv1.*yex_x+adv2.*yex_y)+...
160            reac.*yex+nonlin.reac;
161    end
162
163
164
165    %% Boundary Conditions
166
167    % Drichlet Boundary Condition
168    function DBC=DBCexact(fdiff,x,y)
169        % Evaluate the diffusion function
170        diff = feval(fdiff,x,y);
171        % Drichlet Boundary Condition
172        DBC = 0.5*(1-tanh((2*x-y-0.25)./(sqrt(5*diff))));
173    end
174
175    % Neumann Boundary Condition
176    function NC = NBCexact(mesh,fdiff,x,y)
177        %Neumann Boundary Condition
178        NC = zeros(size(x));
179    end
180
181    %% Set up  parameters function for DG FEM
182
183    function [penalty,kappa]=set_parameter(method,degree)
184
185        global Equation;
186
187        % Superpenalization parameter (In standart b0=1)
188        Equation.b0 = 1;
189        % Choose the basis ( 1:monomials, 2:Dubiner Basis)
190        Equation.base = 2;
191
192        switch method
193            case 1
194              % NIPG
195              Equation.method = 1;
196              kappa = 1;        % type of primal method
197              penalty = 1;      % penalty parameter
198            case 2
199              % SIPG
200              Equation.method = 2;
201              kappa = -1;                          % type of primal method
202              penalty = 3*degree*(degree+1); % penalty parameter
203            case 3
204              % IIPG
205              Equation.method = 3;
206              kappa = 0;                          % type of primal method
207              penalty = 3*degree*(degree+1); % penalty parameter
208
209        end
210
211    end
```

Matlab Code for Matrix Reordering

```
function [N, partitionPoint] = MatrixReorder(M)
  % Authors:    Omer Tari, Murat Manguoglu
  %             June 2013
  %
  % This function reorders a sparse matrix using the largest
  % eigenvector of the Laplacian of the graph corresponding
  % to the matrix. Reordered matrix can be partitioned into
  % 2x2 blocks
  %
  % Input:    M - sparse input matrix
  % Output:   N - resulting reordered matrix
  %           partitionPoint - row and column index of
  %           the partition borders
  %
  % Reference: Tari O, Manguoglu M, ''Revealing the Saddle Point
  % Structure Using the Largest Eigenvector of the Laplacian'',
  % International Conference On Preconditioning Techniques For
  % Scientific And Industrial Applications (19-21 June 2013),
  % Oxford, UK
  %
    % Get size of the Matrix
    [x,y] = size(M);

    theRank = sprank(M);

    if( theRank ~= x || x ~= y )
        return;
    end
    permute = symrcm(M);
    M = M(permute,permute);

    % If not symmetric make symmetric
    isSym = M - M';
    if( nnz(isSym) == 0 )
        isSym = abs(M);
    else
        isSym = abs(M) + abs(M');
    end

    % Get diagonal
    Diagonal = diag(diag(isSym));
    % Get rest
    Rest = isSym - Diagonal;
    % Row , Col and Values of Sparse
    [row,col,values] = find(Rest);
    values = -1;

    RestNew = sparse(row,col,values,x,y);
    summationVector = sum(abs(RestNew'));
```

```
    DiagonalNew = diag(summationVector);
    Laplacian = DiagonalNew + RestNew;

    opts.accuracy = 1e-12;
    [eigenVectors,eigenValues] = eigs(Laplacian,5,'lm',opts);
    theEigenVector = eigenVectors(1:x,1);

    partitionPoint = 0;
    absEigenVector(x) = 0;
    for i = 1:x
        if ( theEigenVector(i) <= 0 )
            absEigenVector(i) = - theEigenVector(i);
            partitionPoint = partitionPoint+1;
        end
    end

    [t,p] = sort(absEigenVector,'descend');
    N = M(p,p);
end
```

References

1. Adjerid, S., Flaherty, J.E., Babuška, I.: A posteriori error estimation for the finite element method-of-lines solution of parabolic problems. Math. Models Methods Appl. Sci. **9**(2), 261–286 (1999)
2. Ainsworth, M.: A posteriori error estimation for discontinuous Galerkin finite element approximation. SIAM J. Numer. Anal. **45**, 1777–1798 (2007)
3. Ainsworth, M., Oden, J.T.: A posteriori error estimation in finite element analysis. Wiley-Interscience (1997)
4. Antonietti, P.F., Süli, E.: Domain decomposition preconditioning for discontinuous Galerkin approximations of convection-diffusion problems. In: Domain Decomposition Methods in Science and Engineering XVIII, *Lecture Notes in Computational Science and Engineering*, vol. 70. Springer Berlin Heidelberg (2009)
5. Arnold, D., Brezzi, F., Cockborn, B., Marini, L.: Unified analysis of discontinuous Galerkin methods for elliptic problems. SIAM J. Numer. Anal. **39**, 1749–1779 (2002)
6. Arnold, D.N.: An interior penalty finite element method with discontinuous elements. SIAM J. Numer. Anal. **19**, 724–760 (1982)
7. Ayuso, B., Marini, L.D.: Discontinuous Galerkin methods for advection-diffusion-reaction problems. SIAM J. Numer. Anal. **47**, 1391–1420 (2009)
8. Babuška, I., , Zlámal, M.: Nonconforming elements in the finite element method with penalty. SIAM J. Numer. Anal. **10**, 45–59 (1973)
9. Babuška, I., Feistauer, M., Šolin, P.: On one approach to a posteriori error estimates for evolution problems solved by the method of lines. Numer. Math. **89**, 225–256 (2001)
10. Babuška, I., Strouboulis, T.: The finite element method and its reliability. Numerical mathematics and scientific computation. Clarendon Press (2001)
11. Barnard, S.T., Pothen, A., Simon, H.: A spectral algorithm for envelope reduction of sparse matrices. Numerical linear algebra with applications **4**, 317–334 (1995)
12. Bassi, F., Rebay, S.: A high order accurate discontinuous finite element method for the numerical solution of the compressible Navier-Stokes equations. J. Comput. Phys. **131**, 267–279 (1997)
13. Bastian, P., Engwer, C., Fahlke, J., Ippisch, O.: An unfitted discontinuous Galerkin method for pore-scale simulations of solute transport. Mathematics and Computers in Simulation **81**(10), 2051–2061 (2011)
14. Baumann, C.E., Oden, J.T.: A discontinuous hp finite element method for convection-diffusion problems. Comput. Methods Appl. Mech. Engrg. **175**, 311–341 (1999)
15. Bause, M.: Stabilized finite element methods with shock-capturing for nonlinear convection-diffusion-reaction models. In: Numerical Mathematics and Advanced Applications 2009, pp. 125–133. Springer Berlin Heidelberg (2010)
16. Bause, M., Schwegler, K.: Analysis of stabilized higher-order finite element approximation of nonstationary and non-linear convection-diffusion-reaction equations. Comput. Methods Appl. Mech. Engrg. **209–212**, 184–196 (2012)

M. Uzunca, *Adaptive Discontinuous Galerkin Methods for Non-linear Reactive Flows*,
Lecture Notes in Geosystems Mathematics and Computing,
DOI 10.1007/978-3-319-30130-3

17. Bause, M., Schwegler, K.: Higher order finite element approximation of systems of convection-diffusion-reaction equations with small diffusion. Journal of Computational and Applied Mathematics **246**, 52–64 (2013)
18. Becker, R., Hansbo, P., Larson, M.G.: Energy norm a posteriori error estimation for discontinuous Galerkin methods. Comput. Methods Appl. Mech. Engrg. **192**, 723–733 (2003)
19. Benzi, M., Golub, G.H., Liesen, J.: Numerical solution of saddle point problems. Acta numerica **14**, 1–137 (2005)
20. Bergam, A., Bernardi, C., Mghazli, Z.: A posteriori analysis of the finite element discretization of some parabolic equations. Math. Comp. **74**, 1117–1138 (2005)
21. Bochev, P.B., Gunzburger, M.D.: Finite element methods of least-squares type. SIAM Review **40**, 789–837 (1998)
22. Bochev, P.B., Gunzburger, M.D.: Least-squares finite element methods. Applied Mathematical Sciences, Vol. 166. Springer (2009)
23. Braess, D., Fraunholz, T., Hoppe, R.H.W.: An equilibrated a posteriori error estimator for the interior penalty discontinuous Galerkin method. SIAM Journal on Numerical Analysis **52**(4), 2121–2136 (2014). DOI 10.1137/130916540
24. Brezzi, F., Manzini, G., Marini, D., Pietra, P., Russo, A.: Discontinuous Galerkin approximations for elliptic problems. Numer. Methods Partial Differential Equations **16**, 365–378 (2000)
25. Bürger, R., Sepùlveda, M., Voitovich, T.: On the Proriol-Koornwinder-dubiner hierarchical orthogonal polynomial basis for the DG-FEM (2009)
26. Cangiani, A., Georgoulis, E.H., Metcalfe, S.: Adaptive discontinuous Galerkin methods for nonstationary convection-diffusion problems. IMA Journal of Numerical Analysis pp. 1–20 (2013)
27. Castillo, P.: Performance of discontinuous Galerkin methods for elliptic PDEs. SIAM J. Sci. Comput. **24**, 524–547 (2012)
28. Castro, C.E., Käser, M., Toro, E.F.: Space–time adaptive numerical methods for geophysical applications. Philosophical Transactions of the Royal Society of London A: Mathematical, Physical and Engineering Sciences **367**(1907), 4613–4631 (2009). DOI 10.1098/rsta.2009.0158
29. Chen, L.: *i*FEM: an innovative finite element methods package in MATLAB. Tech. rep., Department of Mathematics, University of California, Irvine (2008)
30. Chen, Z., Feng, J.: An adaptive finite element algorithm with reliable and efficient error control for linear parabolic problems. Math. Comp. **73**, 1167–1193 (2004)
31. Cockburn, B., Shu, C.W.: The local discontinuous Galerkin method for time-dependent convection-diffusion systems. SIAM J. Numer. Anal. **35**, 2440–2463 (1999)
32. Deng, S., Cai, W.: Analysis and application of an orthogonal nodal basis on triangles for discontinuous spectral element methods. Appl. Num. Anal. Comp. Math **2**, 326–345 (2005)
33. Di Pietro, D.A., Vohralik, M.: A review of recent advances in discretization methods, a posteriori error analysis, and adaptive algorithms for numerical modeling in geosciences. Geosciences Numerical Methods **69**, 701–729 (2014)
34. Dobrev, V.A., Lazarov, R.D., Zikatanov, L.T.: Preconditioning of symmetric interior penalty discontinuous Galerkin FEM for elliptic problems. In: Domain Decomposition Methods in Science and Engineering XVII, *Lecture Notes in Computer Science and Engineering*, vol. 60, pp. 33–44. Springer (2008)
35. Dolejši, V.: Analysis and application of IIPG method to quasilinear nonstationary convection-diffusion problems. J. Comp. Appl. Math. **222**, 251–273 (2008)
36. Dolejši, V.: hp-DGFEM for non-linear convection-diffusion problems. Mathematics and Computers in Simulation **87**, 87–118 (2013)
37. Dolejší, V., Ern, A., Vohralík, M.: A framework for robust a posteriori error control in unsteady nonlinear advection-diffusion problems. SIAM J. Numer. Anal. **51**, 773–793 (2013)
38. Dolejší, V., Feistauer, M., Sobotíková, V.: Analysis of the discontinuous Galerkin method for nonlinear convection-diffusion problems. Comput. Methods Appl. Mech. Engrg. **194**, 2709–2733 (2005)

39. Dörfler, W.: A convergent adaptive algorithm for Poissons equations. SIAM Journal on Numerical Analysis **33**, 1106–1124 (1996)
40. Dougles, J., Dupont, T.: Interior penalty procedures for elliptic and parabolic Galerkin methods. In: R. Glowinski, J.L. Lions (eds.) Computing Methods in Applied Sciences, *Lecture Notes in Phys*, vol. 58, pp. 207–216. Springer Berlin Heidelberg (1976)
41. Dunavant, D.A.: High degree efficient symmetrical Gaussian quadrature rules for the triangle. Internat. J. Numer. Methods Engrg. **21**, 1129–1148 (1985)
42. Epshteyn, Y., Rivière, B.: Estimation of penalty parameters for symmetric interior penalty Galerkin methods. J. Comput. Appl. Math. **206**, 843–872 (2007)
43. Ern, A., Stephansen, A.F., Vohralík, M.: Guaranteed and robust discontinuous Galerkin a posteriori error estimates for convection-diffusion-reaction problems. Journal of Computational and Applied Mathematics **234**, 114–130 (2010)
44. Ern, A., Vohralík, M.: A posteriori error estimation based on potential and flux reconstruction for the heat equation. SIAM J. Numer. Anal. **48**, 198–223 (2010)
45. Fiedler, M.: Algebraic connectivity of graphs. Czechoslovak Mathematical Journal **23**, 298–305 (1973)
46. Fiedler, M.: A property of eigenvectors of nonnegative symmetric matrices and its application to graph theory. Czechoslovak Mathematical Journal **25**, 619–633 (1975)
47. Georgoulis, E., Lakkis, O., Virtanen, J.: A posteriori error control for discontinuous Galerkin methods for parabolic problems. SIAM Journal on Numerical Analysis **49**(2), 427–458 (2011)
48. Georgoulis, E., Loghin, D.: Norm preconditioners for discontinuous Galerkin hp-finite element methods. SIAM Journal on Scientific Computing **30**, 2447–2465 (2008)
49. Gerritsen, M.G., Durlofsky, L.J.: Modeling fluid flow in oil reservoirs. Annual Review of Fluid Mechanics **37**(1), 211–238 (2005). DOI 10.1146/annurev.fluid.37.061903.175748
50. Glassley, W.E., Nitao, J.J., Grant, C.W.: Three-dimensional spatial variability of chemical properties around a monitored waste emplacement tunnel. Journal of Contaminant Hydrology **62–63**, 495–507 (2003)
51. Gopalakrishnan, J., Kanschat, G.: A multilevel discontinuous Galerkin method. Numer. Math **95**, 527–550 (2003)
52. Hauke, G.: A simple subgrid scale stabilized method for the advection-diffusion-reaction equation. Comput. Methods. in Applied Mech. and Eng. **191**, 2925–2947 (2002)
53. Hoppe, R.H.W., Kanschat, G., Warburton, T.: Convergence analysis of an adaptive interior penalty discontinuous Galerkin method. SIAM Journal on Numerical Analysis **47**, 534–550 (2008)
54. Houston, P., Jensen, M., Süli, E.: hp-discontinuous Galerkin finite element methods with least-squares stabilization. J. Sci. Comput. **17**, 3–25 (2002)
55. Houston, P., Schötzau, D., Wihler, T.P.: Energy norm a posteriori error estimation of hp-adaptive discontinuous Galerkin methods for elliptic problems. Mathematical Models and Methods in Applied Sciences **17**, 33–62 (2007)
56. Houston, P., Schwab, C., Süli, E.: Discontinuous hp-finite element methods for advection-diffusion-reaction problems. SIAM J. Numer. Anal. **39**, 2133–2163 (2002)
57. Hughes, T.J.R., Franca, L.P., Hulbert, G.M.: A new finite element formulation for computational fluid dynamics: VIII. the Galerkin/least-squares method for advective-diffusive equations. Comput. Methods Appl. Mech. Engrg. **73**, 173–189 (1989)
58. Jiang, B.N.: The Least-squares Finite Element Method. Theory and Applications in Computational Fluid Dynamics and Electromagnetics. Springer (1998)
59. Johnson, C., Pitkäranta, J.: An analysis of the discontinuous Galerkin method for a scalar hyperbolic equation. Math. Comp. **46**, 1–26 (1986)
60. Karakashian, O.A., Pascal, F.: A posteriori error estimates for a discontinuous Galerkin approximation of second-order elliptic problems. SIAM Journal on Numerical Analysis **41**, 2374–2399 (2003)
61. Karakashian, O.A., Pascal, F.: Convergence of adaptive discontinuous Galerkin approximations of second-order elliptic problems. SIAM Journal on Numerical Analysis **45**, 641–665 (2007)

62. Klieber, W., Rivière, B.: Adaptive simulations of two-phase flow by discontinuous Galerkin methods. Computer Methods in Applied Mechanics and Engineering **196**(1–3), 404–419 (2006)

63. Kunert, G.: A posteriori error estimation for convection dominated problems on anisotropic meshes. Math. Methods Appl. Sci. **26**, 589–617 (2003)

64. Lazarov, R.D., Vassilevski, P.S.: Least-squares streamline diffusion finite element approximations to singularly perturbed convection-diffusion problems. In: L.G. Vulkov, J.J.H. Miller, G.I. Shishkin (eds.) Analytical and Numerical Methods for Singularly Perturbed Problems, pp. 83–94. Nova Science Publishing House (2000)

65. Lesaint, P., Raviert, P.A.: On a finite element for solving the neutron transport equation, mathematical aspects of finite elements in partial differential equations. Math. Res. Center, Univ. of Wisconsin-Madison, Academic Press, New York pp. 89–123 (1974)

66. Leva, P.d.: MULTIPROD TOOLBOX, multiple matrix multiplications, with array expansion enabled. Tech. rep., University of Rome Foro Italico, Rome (2008)

67. Makridakis, C., Nochetto, R.H.: Elliptic reconstruction and a posteriori error estimates for parabolic problems. SIAM J. Numer. Anal. **41**(4), 1585–1594 (2003)

68. Manguoğlu, M., Koyutürk, M., Sameh, A.H., Grama, A.: Weighted matrix ordering and parallel banded preconditioners for iterative linear system solvers. SIAM Journal on Scientific Computing **32**, 1201–1216 (2010)

69. Peraire, J., Persson, P.O.: The compact discontinuous Galerkin (CDG) method for elliptic problems. SIAM J. Sci. Comput. **30**, 1806–1824 (2008)

70. Persson, P., Peraire, J.: Sub-cell shock capturing for discontinuous Galerkin methods (2006)

71. Picasso, M.: Adaptive finite elements for a linear parabolic problem. Comput. Methods Appl. Mech. Engrg. **167**, 223–237 (1998)

72. Proft, J., Riviere, B.: Discontinuous Galerkin methods for convection-diffusion equations for varying and vanishing diffusivity. Int. J. Numer. Anal. Model **6**(4), 533–561 (2009)

73. Reed, W.H., Hill, T.R.: Triangular mesh methods for the neutron transport equation. Tech. Rep. LA-UR-73-479, Los Alomos Scientific Laboratory, Los Alomos, NM (1973)

74. Rivière, B.: Discontinuous Galerkin methods for solving elliptic and parabolic equations, Theory and implementation. SIAM (2008)

75. Rivière, B.: Discontinuous finite element methods for coupled surface-subsurface flow and transport problems. In: IMA Volumes in Mathematics and its Applications: Recent Developments in Discontinuous Galerkin Finite Element Methods for Partial Differential Equations, pp. 259–279. Springer (2013). DOI 10.1007/978-3-319-01818-8_11

76. Rivière, B., Wheeler, M.F.: A posteriori error estimates for a discontinuous Galerkin method applied to elliptic problems. Computers Math. with Applications **46**, 141–163 (2003)

77. Rivière, B., Wheeler, M.F., Girault, V.: Improved energy estimates for interior penalty, constrained and discontinuous Galerkin methods for elliptic problems. Comput. Geosci. **3**, 337–360 (1999)

78. Sangalli, G.: Robust a-posteriori estimator for advection-diffusion-reaction problems. Math. Comp. **77**, 41–70 (2008)

79. Scheibe, T., Yabusaki, S.: Scaling of flow and transport behavior in heterogeneous groundwater systems. Advances in Water Resources **22**(3), 223 – 238 (1998)

80. Schötzau, D., Zhu, L.: A robust a-posteriori error estimator for discontinuous Galerkin methods for convection-diffusion equations. Applied Numerical Mathematics **59**, 2236–2255 (2009)

81. Solin, P., Segeth, K., Dolezel, I.: Higher-Order Finite Element Methods. Chapman & Hall/CRC Press (2003)

82. Solsvik, J., Tangen, S., Jakobsen, H.A.: Evaluation of weighted residual methods for the solution of the pellet equations: The orthogonal collocation, Galerkin, tau and least-squares methods. Computers and Chemical Engineering **58**, 223–259 (2013)

83. Steefel, C.I., DePaolo, D.J., Lichtner, P.C.: Reactive transport modeling: An essential tool and a new research approach for the earth sciences. Earth and Planetary Science Letters **240**(34), 539 – 558 (2005). DOI dx.doi.org/10.1016/j.epsl.2005.09.017

84. Sun, S., Wheeler, M.F.: L2(H1) norm a posteriori error estimation for discontinuous Galerkin approximations of reactive transport problems. Journal of Scientific Computing 22–23(1–3), 501–530 (2005)
85. Tambue, A., Lord, G.J., Geiger, S.: An exponential integrator for advection-dominated reactive transport in heterogeneous porous media. Journal of Computational Physics 229, 3957–3969 (2010)
86. Tarı, O., Manguoğlu, M.: Revealing the saddle point structure using the largest eigenvector of the Laplacian (2013). International Conference On Preconditioning Techniques For Scientific And Industrial Applications (19-21 June, 2013), Oxford, UK
87. Tezduyar, T.E., Park, Y.J.: Discontinuity capturing finite element formulations for nonlinear convection-diffusion-reaction equations. Comp. Meth. Appl. Mech. Eng. 59, 307–325 (1986)
88. Uzunca, M., Karasözen, B., Manguoğlu, M.: Adaptive discontinuous Galerkin methods for non-linear diffusion-convection-reaction equations. Computers and Chemical Engineering 68, 24–37 (2014)
89. Van Slingerland, P., Vuik, C.: Fast Iterative Methods for Discontinuous Galerkin Discretizations for Elliptic PDEs. Reports of the Department of Applied Mathematical Analysis. Delft University of Technology (2010)
90. Van Slingerland, P., Vuik, C.: Fast linear solver for diffusion problems with applications to pressure computation in layered domains. Computational Geosciences pp. 1–14 (2014)
91. Verfürth, R.: A review of a posteriori error estimation and adaptive mesh refinement techniques. Wiley-Teubner Series: Advances in Numerical Mathematics. Wiley-Teubner (1996)
92. Verfürth, R.: A posteriori error estimators for convection-diffusion equations. Numer. Math. 80, 641–663 (1998)
93. Verfürth, R.: Robust a posteriori error estimates for stationary convection-diffusion equations. SIAM Journal on Numerical Analysis 43, 1766–1782 (2005)
94. Verfürth, R.: A posteriori Error Estimation Techniques for Finite Element Methods. Oxford University Press (2013)
95. Wheeler, M.F.: An elliptic collocation-finite element method with interior penalties. SIAM J. Numer. Anal. 15, 152–161 (1978)
96. Yang, J., Chen, Y.: A unified a posteriori error analysis for discontinuous Galerkin approximations of reactive transport equations. Journal of Computational Mathematics 24(3), 425–434 (2006)
97. Yücel, H., Heinkenschloss, M., Karasözen, B.: Distributed optimal control of diffusion-convection-reaction equations using discontinuous Galerkin methods. In: A. Cangiani, R.L. Davidchack, E. Georgoulis, A.N. Gorban, J. Levesley, M.V. Tretyakov (eds.) Numerical Mathematics and Advanced Applications 2011, pp. 389–397. Springer Berlin Heidelberg (2013)
98. Yücel, H., Stoll, M., Benner, P.: Discontinuous Galerkin finite element methods with shock-capturing for nonlinear convection dominated models. Computers and Chemical Engineering 58, 278–287 (2013)

Printed in the United States
by Baker & Taylor Publisher Services